Maria Han Veiga and François Gaston Ged
The Mathematics of Machine Learning

Also of Interest

Mathematics of Deep Learning. An Introduction
Leonid Berlyand, Pierre-Emmanuel Jabin, 2023
ISBN 978-3-11-102431-8, e-ISBN (PDF) 978-3-11-102555-1,
e-ISBN (EPUB) 978-3-11-102580-3

Complex Heterogeneous Systems. Thermodynamics, Information Theory,
Composites, Networks, and Electrochemistry
Markus Schmuck, 2024
ISBN 978-3-11-057953-6, e-ISBN (PDF) 978-3-11-057954-3,
e-ISBN (EPUB) 978-3-11-057952-9

Dynamic Fuzzy Machine Learning
Fanzhang Li, Li Zhang, Zhao Zhang, 2017
ISBN 978-3-11-051870-2, e-ISBN (PDF) 978-3-11-052065-1,
e-ISBN (EPUB) 978-3-11-051875-7

Pattern Recognition. Introduction, Features, Classifiers and Principles
Jürgen Beyerer, Raphael Hagmanns, Daniel Stadler, 2024
ISBN 978-3-11-133919-1, e-ISBN (PDF) 978-3-11-133920-7,
e-ISBN (EPUB) 978-3-11-133941-2

Automatic Complexity. A Computable Measure of Irregularity
Bjørn Kjos-Hanssen, 2024
ISBN 978-3-11-077481-8, e-ISBN (PDF) 978-3-11-077487-0,
e-ISBN (EPUB) 978-3-11-077490-0

Maria Han Veiga and François Gaston Ged

The Mathematics of Machine Learning

—

Lectures on Supervised Methods and Beyond

DE GRUYTER

Mathematics Subject Classification 2020
Primary: 68T01, 68T05, 68T07; Secondary: 68T20, 68W01

Authors

Dr. Maria Han Veiga
Department of Mathematics
Ohio State University
231 W 18th Ave
Columbus, OH 43210
USA
hanveiga.1@osu.edu

Dr. François Gaston Ged
EPFL SB MATH CSFT
MA C1 652 (Bâtiment MA)
Station 8
1015 Lausanne
Switzerland
fged.math@gmail.com

ISBN 978-3-11-128847-5
e-ISBN (PDF) 978-3-11-128899-4
e-ISBN (EPUB) 978-3-11-128981-6

Library of Congress Control Number: 2024931730

Bibliographic information published by the Deutsche Nationalbibliothek
The Deutsche Nationalbibliothek lists this publication in the Deutsche Nationalbibliografie;
detailed bibliographic data are available on the Internet at http://dnb.dnb.de.

Preface

These notes were first developed from a course on the mathematical foundations of machine learning, taught at the University of Michigan in the Fall semester of 2021. The second, third, and nth iterations of these notes have been a work in progress and a collaboration between Maria Han Veiga and François Ged. We write these notes in a way that we believe is helpful for mathematicians to understand the fundamental principles of Machine Learning.

These lecture notes are aimed at being an introduction to machine learning, with a strong focus on the mathematics behind a lot of the algorithms and techniques. Although these notes are suited for a mathematics student, we also want to give the opportunity to do things hands-on, so each chapter finishes with some implementation details related to the topic of discussion.

A special thanks to Prof. Christian Klingenberg and Kathrin Hellmuth for using (an early version of) these notes and for many corrections they provided. We are very interested in receiving corrections, comments, and suggestions via electronic email at hanveiga.1@osu.edu.

https://doi.org/10.1515/9783111288994-201

Contents

Part I: **Introduction and preliminaries**

1 Introduction to machine learning

Machine learning aims at building algorithms that autonomously learn how to perform a task from examples. This definition is rather vague on purpose, but to make it slightly clearer, by "autonomously" we mean that no expert is teaching (or coding by hand) the solution; by "learn" we mean that we have a measure of performance of the algorithm output on the task. In this chapter, we wish to give a general and accessible picture of machine learning through a mathematical formalism. We will introduce the notation, setup, and essential concepts of the field without dwelling on details. The current chapter should provide enough formalism and intuition to make the forthcoming chapters appear natural to the reader.

1.1 Different paradigms

There are three main paradigms in machine learning, which sometimes share similar ideas while having very specific techniques. Namely,
– *Supervised learning* – we have access to labeled examples. For example, the task of spam detection using a dataset of emails, some of which we know are spam, and the others we know are not spam.
– *Unsupervised learning* – examples are not labeled. For example, the dataset is composed of paintings, and the algorithm must group them by guessing which come from the same artist or share the same style.
– *Reinforcement learning* – examples are generated from interacting with the environment. For example, the algorithm controls a drone and learns how to navigate the world by trial and error.

We note that these different types of learning are not mutually exclusive. In this book, we mostly focus on supervised learning; many aspects of our treatment can be transferred to unsupervised learning (Chapter 10) and reinforcement learning (Chapter 11). We hope that the reader is then able to learn autonomously from the literature.

1.2 Supervised learning

In the previous section, we said that supervised learning is learning through *labeled* examples. The problems that can be solved are divided into two types: classification problems, where the goal is prediction of a *categorical* variable, and regression problems, where the prediction takes on *numerical* values that are compared with a distance map.

https://doi.org/10.1515/9783111288994-001

Example 1.2.1. Recognizing handwritten digits is a classification task: each digit belongs to one of the ten classes from "zero" to "nine". If a model reads a 2 instead of a 5, then it is as wrong as if it has read a 1.

Example 1.2.2. Predicting tomorrow's weather (say temperature) is a regression task: if a model predicts a temperature of 24 °C tomorrow in Ann Arbor and it happens to be 27 °C, although not exact, this prediction is more satisfactory than a prediction of 12 °C.

The term *supervised* refers to the fact that the examples used in building the predictor come with labels, that is, learning how to distinguish handwritten digits is done by presenting images *and* the right answers to the algorithm. This is in contrast to *unsupervised learning*, where no labels are provided, and the main goal is to find a *structure* in the data (for example, possible clusters, lower-dimensional representations, etc.).

1.2.1 Setup

The supervised learning setup is as follows:[1]
- An input space $\mathcal{X} \subset \mathbb{R}^{n_{in}}$ with $n_{in} \geq 1$ and an output space $\mathcal{Y} \subset \mathbb{R}^{n_{out}}$ with $n_{out} \geq 1$.
- An unknown mapping $f : \mathcal{X} \to \mathcal{Y}$ we want to approximate.
- A probability distribution D on $\mathcal{X}$.
- A dataset $S = \{(x_i, y_i); i = 1, \ldots, m\}$ such that $f(x_i) = y_i$ for all $i = 1, \ldots, m$ and the x_i are independent and identically distributed (i. i. d.) with common law D.
- A *hypothesis class* $\mathcal{H}$, that is, a set of functions mapping $\mathcal{X}$ to $\mathcal{Y}$, supposedly containing the unknown f (or good approximations of it).
- A loss function $\ell : \mathcal{Y} \times \mathcal{Y} \to \mathbb{R}_+$ such that for $h \in \mathcal{H}$, $\ell(h(x_i), y_i)$ measures the error of prediction of h at x_i from its true label $y_i = f(x_i)$. It is thus standard to assume that $\ell(y, y') = \ell(y', y)$ and $\ell(y, y) = 0$ for all $y, y' \in \mathcal{Y}$.
- A training algorithm $\mathcal{A}$, as defined in the forthcoming Definition 1.2.9.

Remark 1.2.3. The dataset above supposes that the observations are perfect. Often in practice, this is not the case (e. g., temperature measurements can have measurement noise). Such a dataset is said to be *noisy*, and this noise is included in the model so that $y_i = f(x_i) + \epsilon_i$, where $(\epsilon_1, \ldots, \epsilon_n)$ is a random vector, often (but not always) assumed to be Gaussian with mean 0 and independent coordinates. More on that in the next chapter.

Remark 1.2.4. Choosing a parametric model corresponds to choosing a specific hypothesis class $\mathcal{H}$, and hence we will interchangeably use the words *model* and *hypothesis class*. For instance, if $\mathcal{X} \subset \mathbb{R}$, then we could choose $\mathcal{H} = \{h : x \mapsto w_1 \cos(x) + w_2 \sin(x);$

---

1 We provide a short introduction to probability concepts required for the better understanding of this material in Chapter 2.

$w_1, w_2 \in \mathbb{R}$}. The numbers w_1 and w_2 are called the *parameters* of the model. Throughout these notes, we will aggregate them in a parameters vector $w \in \mathbb{R}^P$, where P will usually denote the number of parameters of the model (here $P = 2$). Henceforth, we call an element of $\mathcal{H}$ a *predictor*, and we write f_w instead of h for a function in $\mathcal{H}$ when we want to make the dependency of the predictor on the parameters explicit.

Remark 1.2.5. Choosing a hypothesis class $\mathcal{H}$ is often an engineering choice; we might choose $\mathcal{H}$ according to simplicity, expressiveness, prior knowledge of our problem, etc. In this book, we will see, for example; perceptrons, support vector machines, kernel methods, ensemble methods, neural networks, etc.

Now we can define more concretely classification and regression problems.

Definition 1.2.6. A problem is said to be a classification problem if the labels are categorical or, more formally, if $\mathcal{Y}$ is discrete and $\ell : (y,y') = \mathbb{1}_{\{y \neq y'\}}$. If there are $k \in \mathbb{N}$ classes, then we usually encode them as $\mathcal{Y} = \{1, \ldots, k\}$.

Definition 1.2.7. A regression task is a learning task where $\mathcal{Y}$ takes numerical values, i. e., $\mathcal{Y} \in \mathbb{R}^{n_{out}}$, and predictions are evaluated by a loss function $\ell : \mathcal{Y} \times \mathcal{Y} \to \mathbb{R}_+$.

The loss function in a regression task does more than merely discerning the correctness of a prediction; it also provides a magnitude of error. This is the key difference between classification and regression.

Definition 1.2.8. For a predictor $h \in \mathcal{H}$, the *generalization error* of h is defined by

$$R_D(h) = \mathbb{E}_{x \sim D}[\ell(h(x), f(x))], \tag{1.1}$$

where $x \sim D$ means that x is a random variable with law D.

The goal of supervised learning is to solve the optimization problem

$$\min_{h \in \mathcal{H}} R_D(h). \tag{1.2}$$

However, it is impossible to evaluate $R_D(h)$ without knowing D and f, which are the data distribution and the function we try to approximate. However, we know **some** values of f on **some** samples from D through the dataset $S = \{(x_i, y_i); i = 1, \ldots, m\}$, since $y_i = f(x_i)$ and $x_i \sim D$. Hence, instead of the generalization error, we can minimize another quantity and solve

$$\min_{h \in \mathcal{H}} L(h) = \min_{h \in \mathcal{H}} \frac{1}{m} \sum_{i=1}^{m} \ell(h(x_i), y_i). \tag{1.3}$$

The mapping $L : \mathcal{H} \to \mathbb{R}_+$ is called the *empirical error* (or *training error, empirical/training loss*), as it depends on the dataset. What we described above is the **Empirical Risk Minimization** (ERM) framework, and it assumes that a predictor that minimizes (1.3), which we denote h_*, is close to minimizing (1.1). In other words, even though we may not manage to find h_*, the hope is that a predictor h that performs decently well on the dataset is good at predicting labels for inputs outside the dataset. More concisely, we expect that $L(h)$ small $\Rightarrow R_D(h)$ small. We will see, however, in the forthcoming Section 1.3 that for some h, $L(h)$ can be small – even zero – and $R_D(h)$ very large; this is called *overfitting*. Avoiding overfitting is a topic of both practical and theoretical interest.

1.2.2 Parameters and training

In Remark 1.2.4, we explained that a parametric model defines a hypothesis class; that is, denoting by $\mathcal{W} \subset \mathbb{R}^P$ the parameter space of the model, we have $\mathcal{H} = \{f_w; w \in \mathcal{W}\}$. The parameters w can be trained to search over predictors in $\mathcal{H}$, and hence they are called *trainable parameters*. The training procedure simply refers to the following:

Definition 1.2.9 (Informal). Given a dataset S and a hypothesis class $\mathcal{H} = \{f_w; w \in \mathcal{W}\}$ with parameter space $\mathcal{W} \subset \mathbb{R}^P$, we say that a map $\mathcal{A} = \mathcal{A}_{\mathcal{H},S} : \mathcal{W} \to \mathcal{W}$ is a *training algorithm*.

The implicit idea behind this definition is that the purpose of a training algorithm $\mathcal{A}$ is sending the initial parameters $w_0 \in \mathcal{W}$ to trained parameters $\mathcal{A}(w_0) \in \mathcal{W}$ using the dataset S such that $f_{\mathcal{A}(w_0)}$ performs well at minimizing (1.3). In this context, the scheme used to choose w_0 gives an initial predictor f_{w_0} that does not need to perform well. We can choose w_0 deterministically or randomly, and we call this *the initialization* of the parameters. The training algorithm can itself use randomness.

Often, a parametric model has nontrainable parameters. We make this distinction explicit.

Definition 1.2.10. All parameters of $\mathcal{H}$ and $\mathcal{A}$ that are not modified by $\mathcal{A}$ are called *hyperparameters*.

Changing the hyperparameters corresponds to changing the hypothesis class or/and the algorithm $\mathcal{A}$.

Example 1.2.11. Let $\mathcal{X} = \mathbb{R}$ and $\mathcal{Y} = \mathbb{R}$. Let $\mathcal{H}_k$ be the set of polynomials of degree at most $k \in \mathbb{N}$. Given a dataset $S = \{(x_i, y_i); i = 1, \ldots, m\}$, we can try to learn the task in $\mathcal{H}_1$ if we believe that the relationship between inputs and outputs is linear, i. e., there exist $a, b \in \mathbb{R}$ such that $y_i = ax_i + b$ and, more generally, $f(x) = ax + b$ for all $x \in \mathbb{R}$. A training algorithm $\mathcal{A} : \mathbb{R}^2 \to \mathbb{R}^2$ that finds the best (a, b) does not modify the value of k (equal to 1 here), and hence it is a hyperparameter.

1.3 Model selection

In this section, we present a method that addresses the following question:

> "Given a learning task, how do we choose a good model?"

Indeed, choosing a simple model, i. e., a parametric hypothesis class with few parameters, may result in a predictor that is unable to fit the data, whereas a complex model with too many parameters will fit the data but may not be able to predict reasonable values outside the dataset. These issues are called *underfitting* and *overfitting* and are (slightly) more formally defined below. We will see how splitting the dataset into a training set and a *test set* helps avoiding it.

1.3.1 Underfitting and overfitting

Recall that in the ERM framework, in hope to minimize (1.1), we seek to minimize (1.3).

Definition 1.3.1. Let $h \in \mathcal{H}$ be a predictor. In the ERM framework, the difference between the generalization loss of h and its empirical loss is called the *generalization gap* of h, that is,

$$R_D(h) - L(h) = \mathbb{E}_{x \sim D}[\ell(h(x), f(x))] - \frac{1}{m} \sum_{i=1}^{m} \ell(h(x_i), y_i).$$

The generalization gap is in general inaccessible to the practitioner, as D and f are unknown. However, there are ways we can estimate it by splitting the dataset into two disjoint training and test sets. Indeed, let S_{train} and S_{test} be two disjoint subsets of S. To ease the notation, let us assume that $S_{\text{train}} = \{(x_i, y_i) : i \in \{1, \ldots, m'\}\}$ and $S_{\text{test}} = \{(x_i, y_i) : i \in \{m'+1, \ldots, m\}\}$ for some $m' < m$. Then it suffices to train a model on S_{train} to fit it and to estimate the generalization error $\mathbb{E}_{x \sim D}[\ell(h(x), f(x))]$ by

$$\frac{1}{m - m'} \sum_{i=m'+1}^{m} \ell(h(x_i), y_i).$$

Note that if $m - m'$ tends to infinity, then this becomes the exact generalization error by the law of large numbers (Theorem 2.8.1) since the data samples are assumed to be i. i. d.

Definition 1.3.2 (Underfitting and overfitting).
- We say that *underfitting* occurs when a hypothesis class is too simple to fit the data properly, that is, when $\inf_{h \in \mathcal{H}} L(h)$ is large.
- We say that *overfitting* occurs when a predictor h fits the data well but is too complex to generalize outside the dataset, that is, when $L(h) \approx 0$ but $R_D(h)$ is large.

In general, the training error decreases as we increase the complexity or flexibility of our model (e. g., in polynomial fitting, as we use higher degrees of polynomial functions). The generalization error tends to also decrease initially as complexity increases, but then increases as the model overfits the training set.[2]

---

2 This is not the complete story, as you will further see.

Remark 1.3.3. In Figure 1.1, overfitting occurs for the polynomial of degree 13 because the predictor is too complex and fits the noise in the data, whereas the polynomial of degree 2 is too simple to capture enough structure in the data. Degree 4 is reasonably close to fitting the data. Note that even without noise, an overly complex model can feature overfitting: e. g., for a dataset of n points on a line, a polynomial of degree $n + k$ can perfectly fit all the datapoints and look very different from a line.

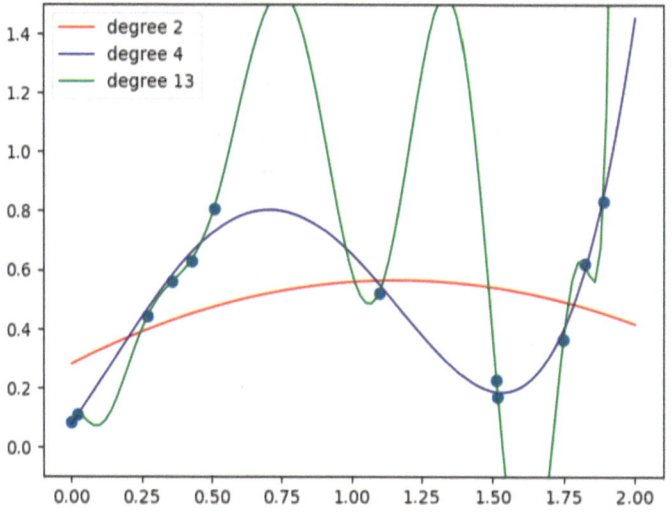

Figure 1.1: Polynomial regression with polynomials of degrees 2, 4, 13.

1.3.2 Validation set and cross-validation

On a given task, to assess the efficacy of a model $\mathcal{H}$, after having split the dataset into a training set and a test set, we can simply look at the test error, as this is an estimate for the generalization loss. Thanks to this splitting, we have a way to detect overfitting, for example, if the training error is very low, but the test error is very high. If our model overfits, then we can simply change to another model. However, that is a naive way of choosing a model, which can lead to **overfitting the test set**.

Indeed, suppose that for a given task, you have the choice between several models $\mathcal{H}^{(1)}, \ldots, \mathcal{H}^{(n)}$ and that you have no *a priori* reason to favor one over the others. How do we select the best model? Suppose that we train all of them on the training set and compare the predictors thus obtained on the test set. We then select the predictor that had the lower test error. By proceeding that way we expose ourselves to selecting a predictor that, **by chance**, overfits S_{test}. Note that the predictor does not have to be trained on S_{test} to overfit it. To clearly see why, suppose that the number of models $n \to \infty$. Then we can convince ourselves that it becomes more and more likely that one of the predictors is such that $h(x_i) \approx y_i$ for all $(x_i, y_i) \in S_{test}$. This means that we cannot assess whether the chosen hypothesis class is well chosen.

Validation set

One way to deal with that issue is to split the dataset into three disjoint subsets:

-   the training set S_{train},
-   the validation set S_{val}, and
-   the test set S_{test}.

Now the procedure becomes the following: train the n models on S_{train}, compare their performances on S_{val}, select the best model and retrain it on $S_{train} \cup S_{val}$, and then assess its performance on S_{test}.

One analogy is to view the training set as textbook lectures and examples from which we learn a new concept, and we encountered some confusing topics that have multiple possible interpretations. The validation set entails practice problems and exams of previous years to help us choose the best interpretation, and the test set is the final exam of the course.

Cross-validation

Although this procedure seems satisfactory, it can be greatly improved. Indeed, note that for each class $\mathcal{H}^{(i)}$, we select **only one** predictor $h^{(i)}$ to assess how good a choice of $\mathcal{H}^{(i)}$ is. This in turn makes the randomness of the finite sampling and splitting of the dataset too important in the model selection; ideally, we want to choose the best model for a distribution D and a labeling function f, not the best model for a given dataset split $S_{train}, S_{val}, S_{set}$.

To solve this issue, we can use *cross-validation*. Let $\mathcal{H}$ be fixed, and let us use cross-validation to assess how good a choice of it is for a given task. Let split the dataset $S = \{(x_i, y_i) : i \in \{1, \ldots, m\}\}$ into a training set S_{train} and test set S_{test} as before. We now randomly partition the dataset $S_{train} = \{(x_i, y_i) : i \in \{1, \ldots, m'\}\}$ into k disjoint and covering subsets $S_1, \ldots, S_k$ of roughly the same size. For $i = 1, \ldots, k$, we call S_i the *ith fold*. We denote by m_i the size of the *i*th fold such that $\sum_{i=1}^{k} m_i = m'$. We now proceed as follows: for all $i \in \{1, \ldots, k\}$, we train the model on

$$\bigcup_{\substack{j=1 \\ j \neq i}}^{k} S_j$$

and denote by $h_i \in \mathcal{H}$ the predictor thus obtained. We evaluate the performance of h_i on the *i*th fold that are the only samples from the training dataset that the predictor has not trained on, that is, we define

$$L_i(h_i) := \frac{1}{m_i} \sum_{(x,y) \in S_i} \ell(h_i(x),y).$$

We now have a collection of k predictors belonging to $\mathcal{H}$, each trained on a different subset of the training dataset, and for each, we have an estimate of their generalization error computed on the corresponding ith fold they have not seen. We now evaluate the quality of the model $\mathcal{H}$ for the task by

$$CV(\mathcal{H}) := \frac{1}{k} \sum_{i=1}^{k} L_i(h_i).$$

Coming back to our initial question of choosing the best model among $\mathcal{H}^{(1)}, \ldots, \mathcal{H}^{(n)}$, we can simply choose the one that minimizes the cross-validation error, that is, the smallest $CV(\mathcal{H}^{(i)})$. Then we can retrain that model on the whole training dataset S_{train} and estimate its generalization error on S_{test}.

Choosing k

Recall that k is the number of folds used during cross-validation (CV). What value of k should we choose?

We can ask ourselves, what is:

- The influence of k when estimating the expected generalization error?
- The influence of k on the size of the training set and therefore attained approximators $h_1, \ldots, h_k$?
- The computational complexity of the training algorithm for different k?

Consider a dataset of fixed size m. On the one extreme, if we choose $k = m'$, then the k-fold CV becomes the *leave-one-out cross-validation (LOO-CV)*. With respect to the estimator of the generalization error, most experiments show that there is a decreasing or constant variance of the generalization error estimator with increasing k.[3] Also, note that this creates the largest number of predictors $h_1, \ldots, h_k$, which might be very similar to one another, with training sets of size $m'' = m' - 1$ (after leaving one out). Furthermore, it can be computationally quite expensive, since it requires solving m' slightly smaller $(m' - 1)$-sized subproblems of the same type.

On the other extreme, a small k (such as $k = 2$) provides an estimator for the generalization error with a higher variance and higher bias. This is due to the fact that the estimators are trained on distinct and smaller datasets ($m'' = m'/2$). However, the computational complexity of the training algorithm is small as well (as we only train h_1 and h_2).

---

3 With some exceptions on the stability of the algorithm.

In practice, the number of folds k will depend on the dataset size, as we must balance the training dataset size $m'' = (k-1)m'/k$ and the computational complexity of training k models. For example, Figure 1.2 shows a learning curve for a regressor, plotting the performance of an estimator h using a 5-fold CV. In this case, if we have a dataset of 200 points, then a 5-fold CV would generate training sets of size $m'' = 160$, which would reach an error that is fairly close to if the full $m' = 200$ were used and the trained predictors would not suffer from much more bias. On the other hand, if we had a dataset set of $m' = 50$ points, then using a 5-fold CV would lead to training sets of size $m'' = 40$, leading to a substantially increased error of the predictor. Thus the precise trade-off is problem dependent and dataset size dependent. As a compromise among bias, variance, and computational cost, k of 5 or 10 are commonly used choices for many applied problems.

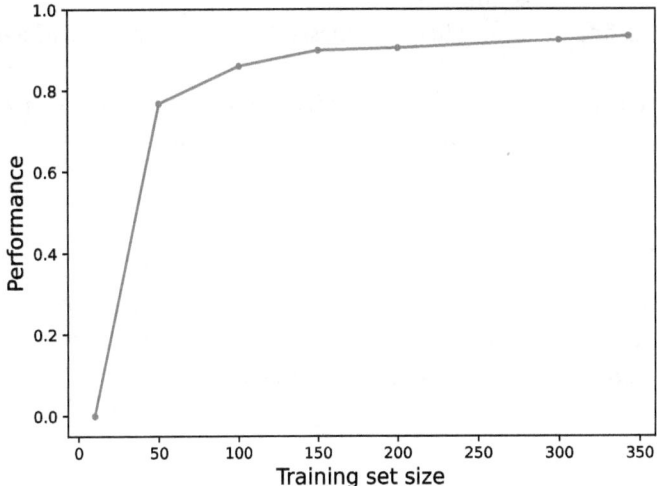

Figure 1.2: An example of a learning curve for a regression on a given task: a plot of the performance (best at 1) versus the size of the training set, considering a 5-fold CV.

1.4 No-free-lunch theorem

The so-called *no-free-lunch theorem* is often talked about informally, one of the reasons being that many similar – but not equivalent – versions of the theorem exist in the literature. The theorem is often stated as follows:

> "All optimization algorithms perform equally well when their performance is averaged across all possible problems."

The term "averaged" here does not have a formal meaning, but using the machine learning formalism we introduced in this chapter, the no-free-lunch theorem can be understood as follows: without any prior knowledge on the data distribution D and the labeling function f, there is no way to guess whether a pair of hypothesis class and training algorithm $(\mathcal{H}_1, \mathcal{A}_1)$ is likely to perform better or worse than another pair $(\mathcal{H}_2, \mathcal{A}_2)$.

It means that there is no single best machine learning algorithm across all possible prediction problems. This in turn motivates the development of many different types of models to cover the wide variety of data that appears in the real world.

1.5 ML pipeline in practice

For most machine learning problems (supervised), the pipeline will be similar:
- Given a dataset, split it into training, validation, and test sets.
- Choose a hypothesis class (or several) $\mathcal{H}$.
- Define a metric of error, i. e., a loss function ℓ.
- Use cross-validation to find the best suited hypothesis class $\mathcal{H}$ in a set of possible hypothesis classes.
- Once the best suited hypothesis class $\mathcal{H}$ has been chosen, train a prediction in $\mathcal{H}$ and valuate the performance on the test set to judge the generalization error.

1.6 Implementation details

In this section, we provide a Python code snippet for cross-validation to compare two hypothesis classes (two types of linear regression; see Chapter 5) using scikit learn. The code can also be found on github: https://github.com/hanveiga/tmml.

Listing 1.1: Cross-validation for linear model using scikit learn

```
1 import numpy as np
2 from sklearn.model_selection import train_test_split,
       KFold
3 from sklearn import datasets, linear_model
4 from sklearn.metrics import mean_squared_error
5
6 # Load generic dataset for regression
7 X, y = datasets.load_diabetes(return_X_y=True)
8
9 # Create hold-out test set
10 X_train, X_test, y_train, y_test = train_test_split(X, y,
       test_size=0.2, random_state=0)
11
12 # Create two hypothesis classes
13 hypothesis_classes = { "Model1": linear_model.Lasso,\
```

```
14                        "Model2": linear_model.Ridge}
15
16 hypothesis_performance = {}
17
18 kf = KFold(n_splits=5)
19
20 for key in hypothesis_classes.keys():
21   fold_performances = []
22   for train, val in kf.split(X_train):
23     # Create linear regression model
24     regressor = hypothesis_classes[key]()
25
26     # Train the model using the K-1 folds
27     regressor.fit(X_train[train,:], y_train[train])
28
29     # Evaluate performance on K-th fold
30     y_pred = regressor.predict(X_train[val,:])
31
32     # Measure loss with mean squared error (MSE)
33     fold_performances.append(mean_squared_error(y_train[
       val], y_pred))
34
35   hypothesis_performance[key] = np.mean(fold_performances)
36
37 # Choose model with lowest MSE and retrain
38 best_model_key = min(hypothesis_performance, key=
       hypothesis_performance.get)
39 best_regressor = hypothesis_classes[best_model_key]()
40 best_regressor.fit(X_train, y_train)
41 y_test_pred = best_regressor.predict(X_test)
42 print(f"MSE: {mean_squared_error(y_test, y_test_pred)}")
```

2 Probability review

In our journey to understand machine learning, we will encounter several sources of randomness, such as those coming from the collected data, which are usually random observations (i. e., samples) of some unknown probability distribution, the initial parameters of the model we will use to make predictions, or the intrinsic randomness of some training algorithms. To build a solid theory, we need some knowledge of probability theory, and this is what this chapter is about.

2.1 Motivation

Let us consider the task of *binary classification*, where we wish to learn a mapping from inputs $x \in \mathcal{X}$ (also called *features*) to outputs $\mathcal{Y} = \{-1, +1\}$. We can formalize the problem as a function approximation problem. Given a labeled training set $\{(x_i, y_i); i = 1, \ldots, m\}$, we assume that there is some unknown function $f : \mathcal{X} \rightarrow \mathcal{Y}$ such that $f(x_i) = y_i$ for all $i = 1, \ldots, m$, and the goal of learning is to approximate the function f by a function $\hat{f} : \mathcal{X} \rightarrow \{-1, 1\}$. Then, for any input $x \in \mathcal{X}$, we can make a prediction of its label using $\hat{y} = \hat{f}(x)$. This is called a *discriminative model*, as its goal is to discriminate between different classes.

We can think of spam detection. In this case, x is an email (or representation of an email); −1 denotes spam, and +1 denotes not spam.

However, to set a milder decision rule, we might prefer to estimate the probability that the email x is a spam and only warn the user that the email is potentially a spam if this probability is larger than some chosen threshold. Having a probability estimate of class membership is even more important when the number of classes is larger than two.

Instead of discriminating whether x belongs to some class y, we might want to create an object x that belongs to a given class y. This is the goal of a *generative model*, which tries to learn the conditional probability $p(x|y)$ instead.

Besides the possibility of making stochastic predictors and stochastic generators, other sources of randomness are more broadly encountered when practicing machine learning. Namely, it is commonly assumed that the dataset is randomly generated by some unknown probability distribution. On the other hand, by using parametric models we need to set a value for the initial parameters before training, and it is common to initialize them at random values. Finally, the training procedure (i. e., the update of the parameters to fit the data) itself can include some randomness. Hence a good understanding of machine learning requires some basic knowledge of probability theory.

https://doi.org/10.1515/9783111288994-002

2.2 Probability space

Probability theory is a mathematical framework that allows us to reason about phenomena or experiments under uncertainty. A probabilistic model is a mathematical model of a probabilistic experiment that satisfies the axioms of probability theory and allows us to calculate probabilities as well as to reason about the outcomes of an experiment.

Definition 2.2.1. A *probability space* is a triplet $(\Omega, \mathcal{F}, P)$, where Ω is an arbitrary nonempty set, $\mathcal{F}$ is a σ-field of subsets of Ω, and P is a measure on $\mathcal{F}$ such that

$$P(\Omega) = 1 \quad \text{and} \quad P\left(\bigcup_{n \geq 1} E_n\right) = \sum_{n \geq 1} P(E_n)$$

for all pairwise disjoint sets $E_n \in \mathcal{F}, n \geq 1$. The measure P is called a probability measure (or in short, probability).

This means that
- Ω is the set of all possible outcomes and is called the *sample space*.
- $\mathcal{F}$ is a collection of subsets of Ω that is a σ-field.[1]
- The probability P maps elements of $\mathcal{F}$ (subsets of Ω) onto the real interval $[0,1]$.
- An element $A \in \mathcal{F}$ is called an event, and $P(A) \in [0,1]$ is the probability that A occurs. See Figure 2.1 for a schematic.

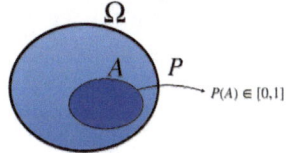

Figure 2.1: Schematic of sample space Ω, event A, and probability $P(A)$ of A.

Example 2.2.2. Suppose we toss a fair coin twice and observe the outcome (the two tosses are independent). We have
- $\Omega = \{HH, TT, HT, TH\}$.
- $\mathcal{F} = \{\{\}\{HH\}, \{TT\}, \{HT\}, \{TH\}, \{HH, TT\}, \{HH, HT\}, \dots, \Omega\}, |\mathcal{F}| = 2^{|\Omega|} = 2^4 = 16$.

Events that depend on the outcome of the experiment can be written as elements of the σ-field $\mathcal{F}$. For example, the event "obtaining exactly one head in those two tosses" is the

---

1 It must contain Ω, and it must be closed by complementation and by countable unions to be a σ-field; we will not manipulate σ-fields in this book, and therefore we do not dwell on it to keep the focus on the necessary concepts.

element $\{HT, TH\}$. Probabilities of events in $\mathcal{F}$ are assigned by the probability measure P. We have, for example, $P(\{\}) = 0$ (always true for any probability measure by definition), and because all outcomes are equally likely (the coin is fair), we have

$$P(\text{"obtaining heads exactly once"}) = \frac{\#\ \text{outcomes with heads exactly once}}{\#\ \text{possible outcomes}} = \frac{2}{4} = \frac{1}{2}.$$

2.3 Independence and conditioning

We say that two events $A, B \in \mathcal{F}$ are independent if the occurrence or nonoccurrence of one does not affect the probability assigned to the other. More formally, we can state the following:

– A and B are independent if $P(A \cap B) = P(A)P(B)$.
– The events in a set $\{A_s \mid s \in S\} \subset \mathcal{F}$ are independent if for every finite subset $S_0 \subset S$, we have

$$P\left(\bigcap_{s \in S_0} A_s\right) = \prod_{s \in S_0} P(A_s). \tag{2.1}$$

To be more precise, the last point is the property of *mutual independence* for a family of events. It is strictly stronger than the property of pairwise independence that A_s and $A_{s'}$ are independent for all $s \neq s'$ in S.

Example 2.3.1. I tossed a fair coin 8 times (tosses are assumed independent), and I obtained head 8 times. What is the probability that I get tails in my next throw?

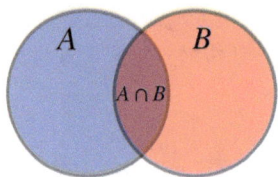

Let $B \in \mathcal{F}$ with $P(B) > 0$. For every event $A \in \mathcal{F}$, the conditional probability that A occurs conditionally given that B occurs is denoted by $P(A|B)$ and defined by

$$P(A|B) = \frac{P(A \cap B)}{P(B)}. \tag{2.2}$$

Example 2.3.2. Given I threw a fair dice and I got an even number, what is the probability it was 2? What about 3?

– A: $\{\text{"Getting 2"}\}$.
– B: $\{\text{"Getting even"}\}$.

- $A \cap B$: {"Getting 2"} and {"Getting even"} = {"Getting 2"},

$$P(A|B) = \frac{P(A \cap B)}{P(B)} = \frac{1/6}{1/2} = 1/3.$$

The following properties can also be proven:
- If $P(B) > 0$ and $\{A_i\}_{i \geq 1}$ are *pairwise* disjoint events, then

$$P\left(\bigcup_{i=1}^{\infty} A_i | B\right) = \sum_{i=1}^{\infty} P(A_i | B). \tag{2.3}$$

- If $B \in \mathcal{F}$ is such that $P(B) > 0$ and $P(B^c) > 0$, where B^c denotes the complement $B^c = \Omega \setminus B$, then

$$P(A) = P(A|B)P(B) + P(A|B^c)P(B^c) \tag{2.4}$$

for all $A \in \mathcal{F}$.
- If $(B_i)_{i \geq 1} \subset \mathcal{F}$ form a partition of Ω (i. e., they are pairwise disjoint and cover Ω) and $P(B_i) > 0$ for all $i \geq 1$, then for all $A \in \mathcal{F}$,

$$P(A) = \sum_{i=1}^{\infty} P(A|B_i)P(B_i). \tag{2.5}$$

2.4 Random variables

A real random variable provides us with a numerical value that is dependent on the outcome of an experiment. It is a convenient way to express the elements of Ω as numbers rather than abstract elements of sets. Throughout this book, we will only consider **real** random variables or **multivariate real** random variables, that is to say, random variables with values in $\mathbb{R}$ or $\mathbb{R}^d$ for $d \geq 2$.

Definition 2.4.1. Let $\mathcal{G}$ be a σ-field on $\mathbb{R}$. A real random variable X is a function $X : \Omega \rightarrow \mathbb{R}$ such that for all $B \in \mathcal{G}$, $\{\omega \in \Omega : X(\omega) \in B\} \in \mathcal{F}$.

Remark 2.4.2. Note that defining a real random variable makes sense only given a probability space $(\Omega, \mathcal{F}, P)$ and a σ-field $\mathcal{G}$ on $\mathbb{R}$. If not specified, it is common to assume that $\mathcal{G}$ is the Borel σ-field on $\mathbb{R}$. (The Borel σ-field is generated by all subintervals of the form $(a, b]$, $a, b \in \mathbb{R}$.)

The event $\{\omega \in \Omega : X(\omega) \in B\}$ in the definition is simply the preimage of B by X, also denoted by $X^{-1}(B)$.

The above definition naturally extends to $X : \Omega \mapsto \mathbb{R}^d$ for all $d \geq 2$.

Example 2.4.3. Toss independently n fair coins and observe the resulting sequence. The state space consists of the set of all 2^n possible coin sequences. Let our random variable

X be the number of heads.[2] For $k \geq 0$, what is $P(X = k)$? That is, in n coin tosses, what is the probability we obtain heads exactly k times?

Example 2.4.4. Toss two independent and fair coins. Our random variable X is the number of heads.

- $X(\text{"}HH\text{"}) = 2.$
- $X(\text{"}HT\text{"}) = 1.$
- $X(\text{"}TH\text{"}) = 1.$
- $X(\text{"}TT\text{"}) = 0.$

Perhaps in more practical terms, a real random variable transforms an element ω from the original sample space (which could be abstract or difficult to work with directly) into a numerical quantity $X(\omega)$ (real number) that is more convenient or tangible to work with (e. g., quantities that we may measure in a laboratory).

Once we are mapped by X from Ω to $\mathbb{R}$, we may choose to view $\mathbb{R}$ as the new "sample space" and create a new probability triple for itself. Then on this new measurable space $(\mathbb{R}, \mathcal{B})$, we can assign probabilities to the events in $\mathcal{B}$, and we call that the *probability distribution* (or *probability law*) of the random variable X, denoted by P_X. The new probability space associated with the random variable X is then $(\mathbb{R}, \mathcal{B}, P_X)$. We can also map from $\mathbb{R}$ back to the original sample space via X^{-1} (note that this is the "preimage" inverse, not the usual inverse function). Hence for some event B in the new σ-field $\mathcal{B}$, we can write

$$P_X(B) = P(X^{-1}(B)) = P(\{\omega : X(\omega) \in B\}). \tag{2.6}$$

2.5 Discrete random variables

2.5.1 Definition

A *discrete* random variable is one whose range $X(\Omega)$ (i. e., the set of values it can take) is countable. The *probability mass function (PMF)* of a discrete random variable is defined as

$$p_X(x) = P(X = x), \quad x \in \mathbb{R}, \tag{2.7}$$

and, in particular,

$$\sum_{x \in \mathbb{R}} p_X(x) = 1. \tag{2.8}$$

---

2 Convince yourself that X is indeed a random variable.

(In the above sum, only countably many terms are non-null). With a slight abuse of language, we say that a random variable X has the distribution, or law, p_X.[3]

2.5.2 Examples of discrete random variables

- $(X \sim \mathcal{U}(\{a, \ldots, b\}))$ Discrete uniform with integer parameters $a < b$ (e. g., throwing a fair dice):

$$p_X(x) = \frac{1}{b - a + 1} \quad \text{for } x \in \{a, \ldots, b\} \quad \text{and} \quad p_X(x) = 0 \text{ otherwise.}$$

- $(X \sim \text{Ber}(p))$ Bernoulli with parameter $0 \le p \le 1$ (e. g., yes or no):

$$p_X(k) = p^k (1 - p)^{1-k} \quad \text{for } k \in \{0, 1\} \quad \text{and} \quad p_X(k) = 0 \text{ otherwise.}$$

- $(X \sim \text{Bin}(n, p))$ Binomial with parameters $n \in \mathbb{N}$ and $p \in [0, 1]$ (e. g., the number of heads after n independent coin tosses with a biased coin):

$$p_X(k) = \binom{n}{k} p^k (1 - p)^{n-k} \quad \text{for } k = 0, 1, \ldots, n \quad \text{and} \quad p_X(k) = 0 \text{ otherwise.}$$

- $(X \sim \text{Pois}(\lambda))$ Poisson with parameter $\lambda > 0$:

$$p_X(k) = e^{-\lambda} \lambda^k / k! \quad \text{for } k = 0, 1, \ldots \quad \text{and} \quad p_X(k) = 0 \text{ otherwise.}$$

The Poisson distribution is typically used to model the number of times an event occurs in a unit amount of time when these occurrences are thought to be independent and when the rate of occurrence is λ, that is, the average number of occurrences per unit of time is λ (e. g., earthquakes, victory of France in the soccer world cup[4]).

2.5.3 Multiple random variables (marginal, joint, conditional, independence)

Consider two discrete random variables X and Y associated with the same experiment. The probability law of each one of them is described by its respective PMF p_X or p_Y, called the *marginal PMFs* of the couple (X, Y). The marginal PMFs do not provide any information on possible relations between these two random variables.

The statistical properties of two random variables X and Y are captured by their *joint PMF*:

$$p_{X,Y}(x, y) = P(X = x, Y = y). \tag{2.9}$$

---

3 Even though technically its law is P_X defined in the previous section (that is because P_X characterizes p_X).

4 In this case, $\lambda = 2/25$ choosing a world cup as the unit of time.

Example 2.5.1. Toss a fair coin and let $X = 1$ if the result is heads and $X = 0$ if it is tails. Let $Y = X$ and $Y' = 1 - X$. Show that (X, Y) and (X, Y') have the same marginal PMFs but not the same joint PMFs.

We may also concatenate multiple random variables together into a random vector $X = (X_1, \ldots, X_n)$ and still use the notation $p_X(x)$ but with the understanding that it is the joint PMF (in this case, $x = (x_1, \ldots, x_n)$). From the joint PMF we can recover the marginals via

$$p_X(x) = \sum_y p_{X,Y}(x, y), \quad p_Y(y) = \sum_x p_{X,Y}(x, y). \tag{2.10}$$

The *conditional PMF* of X given Y is

$$p_{X|Y}(x|y) = P(X = x | Y = y), \tag{2.11}$$

which is well defined whenever $p_Y(y) > 0$. Using the definition of conditional probabilities, we obtain

$$p_{X|Y}(x|y) = \frac{p_{X,Y}(x, y)}{p_Y(y)}. \tag{2.12}$$

Visually, if we fix y, then the conditional PMF $p_{X|Y}(x|y)$ can be viewed as a "slice" of the joint PMF $p_{X,Y}$, normalized so that it sums to one.

Random variables X and Y are said to be *independent* if for all $x, y \in \mathbb{R}$,

$$p_{X,Y}(x, y) = p_X(x) p_Y(y) \tag{2.13}$$

or, equivalently, for all $x, y \in \mathbb{R}$ such that $p_Y(y) > 0$,

$$p_{X|Y}(x|y) = p_X(x). \tag{2.14}$$

Furthermore, if X and Y are independent, then functions of these random variables $h(X)$ and $g(Y)$ are also independent, provided that $h(X)$ and $g(Y)$ are random variables. (We indeed need to check that they satisfy Definition 2.4.1; for example, it is always the case where h and g are continuous).

2.5.4 Sequence of discrete random variables

We just saw how the joint PMF encodes the distribution of a couple of discrete random variables. Sometimes, we will want to consider more than two random variables at the same time. A *sequence of random variables* $(X_i)_{i \geq 1}$ is a sequence such that X_i is a random variable for all $i \geq 1$.

We say that the random variables in $(X_i)_{i \geq 1}$ are independent if for all $k \geq 1$ and all $i_1, \ldots, i_k \in \mathbb{N}$ pairwise distinct, we have

$$p_{(X_{i_1}, \ldots, X_{i_k})}(x_1, \ldots, x_k) = \prod_{\ell=1}^{k} p_{X_{i_\ell}}(x_\ell) \quad \forall x_1, \ldots, x_k \in \mathbb{R}.$$

This is sometimes called *mutual independence* to stress that it is strictly stronger than *pairwise independence*, the latter being the weaker property that for all $i \neq j$, the two random variables X_i and X_j are independent.

Example 2.5.2. Let X be a random variable on $\{0, 1\}$ with the following law: $P(X = 1) = 1/2 = P(X = 0)$. Let Y have the same law and be independent from X. Let Z be a random variable such that $Z = X$ if $Y = 1$ and $Z = 1 - X$ if $Y = 0$. It can be shown that the triplet (X, Y, Z) is composed of pairwise independent variables but not mutually independent.

2.6 Continuous random variables

Most of the properties and concepts for continuous random variables will be the same or analogous to its discrete counterpart (by swapping summation with integration).

2.6.1 Definitions

When X takes real continuous values, it is more natural to specify the probability of X being inside some interval $\mathbb{P}(a \leq X \leq b)$, $a < b$. By convention we specify $\mathbb{P}(X \leq x)$ for all $x \in \mathbb{R}$, which is known as the *cumulative distribution function* (CDF) of X, denoted by $F_X(x)$.

A *continuous* (real) random variable is one that has a *probability density function* (PDF) $f_X(x)$ such that

$$F_X(x) = P(X \leq x) = \int_{-\infty}^{x} f_X(t)\, dt. \tag{2.15}$$

Conversely, if the CDF is differentiable (not always true), then

$$f_X(x) = \frac{\partial F_X(x)}{\partial x}. \tag{2.16}$$

Since the CDF is increasing, we have $f_X(x) \geq 0$, and since $\lim_{x \to +\infty} F_X(x) = 1$, we have

$$\int_{-\infty}^{+\infty} f_X(t)\, dt = 1. \tag{2.17}$$

Using the PDF of a continuous random variable, we can compute the probability of various subsets of the real line:

$$P(a < X < b) = P(a \leq X \leq b) = \int_a^b f_X(t)\, dt, \tag{2.18}$$

$$P(X \in B) = \int_B f_X(x)\, dx. \tag{2.19}$$

Remark 2.6.1. From measure theory, for the last equation to make sense, we need B to be *Lebesgue measurable*. Since it is the case of all Borel sets, we can use this formula for all B that can be constructed from intervals. We will always work with such measurable sets throughout the class without recalling.

Remark 2.6.2. Any random variable can be decomposed into continuous and singular parts (the latter does not need to be discrete but can be). For example, $X = 0$ with probability 1/2 and $X = U \sim \mathcal{U}(0,1)$ (the uniform distribution on $(0,1)$) with probability 1/2; then X is neither continuous nor discrete.

2.6.2 Some examples of continuous random variables

– $(X \sim \mathcal{U}(a,b))$ Uniform with parameters a and b (and $a < b$):

$$f_X(x) = 1/(b-a) \text{ for } x \in [a,b].$$

The probability law of a uniform random variable on $[a,b]$ is the Lebesgue measure on $[a,b]$ divided by $b-a$.
– $(X \sim \text{Exp}(\lambda))$ Exponential with $\lambda > 0$:

$$f_X(x) = \lambda e^{-\lambda x}.$$

The exponential distribution is memoryless, that is,

$$\mathbb{P}(X > x + t | X > x) = \mathbb{P}(X > t)$$

for all $x, t > 0$.
– $(X \sim \mathcal{N}(\mu, \sigma^2))$ Normal (Gaussian) with mean $\mu \in \mathbb{R}$ and standard deviation $\sigma > 0$:

$$f_X(x) = \frac{1}{\sigma\sqrt{2\pi}} e^{-\frac{(x-\mu)^2}{2\sigma^2}}. \tag{2.20}$$

The distribution $\mathcal{N}(0,1)$ is called the *standard normal* or *standard Gaussian* distribution.

2.6.3 Multiple random variables (marginal, joint, conditional, independence)

The joint CDF of X and Y is

$$F_{X,Y}(x,y) = P(X \le x, Y \le y) = \int_{-\infty}^{x} \int_{-\infty}^{y} f_{X,Y}(u,v)\, du\, dv, \qquad (2.21)$$

and $f_{X,Y}$ is called the joint PDF. If the CDF is differentiable (not always true), then

$$\frac{\partial^2 F_{X,Y}}{\partial x \partial y}(x,y) = f_{X,Y}(x,y). \qquad (2.22)$$

Similarly to the univariate case, we can compute the probability of an event B:

$$P((X,Y) \in B) = \int_B f_{X,Y}(x,y)\, dx\, dy. \qquad (2.23)$$

Then the marginal PDF of X is

$$f_X(x) = \int_{-\infty}^{\infty} f_{X,Y}(x,y)\, dy. \qquad (2.24)$$

The conditional CDF of X given Y is

$$F_{X|Y}(x|y) = \int_{-\infty}^{x} \frac{f_{X,Y}(u,y)}{f_Y(y)}\, du, \qquad (2.25)$$

where f_Y is the marginal PDF of Y, and we assumed $f_Y(y) > 0$. Then the conditional PDF is

$$f_{X|Y}(x|y) = \frac{f_{X,Y}(x,y)}{f_Y(y)}. \qquad (2.26)$$

We say that X and Y are independent if their joint CDF, equivalently joint PDF, can be factored:

$$F_{X,Y}(x,y) = F_X(x)F_Y(y), \qquad (2.27)$$
$$f_{X,Y}(x,y) = f_X(x)f_Y(y) \quad \forall x, y \in \mathbb{R}. \qquad (2.28)$$

Equivalently, for all $x, y \in \mathbb{R}$ such that $f_Y(y) > 0$,

$$F_{X|Y}(x|y) = F_X(x), \qquad (2.29)$$
$$f_{X|Y}(x|y) = f_X(x). \qquad (2.30)$$

2.6.4 Sequence of continuous random variables

Sequences of continuous random variables are defined identically to sequences of discrete random variables. The mutual independence property translates similarly with the PDFs and CDFs as follows: we say that a sequence of continuous random variables are mutually independent if for all $k \geq 1$ and all pairwise distinct $i_1, \ldots, i_k \in \mathbb{N}$, we have

$$F_{X_{i_1}, \ldots, X_{i_k}}(x_1, \ldots, x_k) = \prod_{\ell=1}^{k} F_{X_{i_\ell}}(x_\ell)$$

or, equivalently,

$$f_{X_{i_1}, \ldots, X_{i_k}}(x_1, \ldots, x_k) = \prod_{\ell=1}^{k} F_f X_{i_\ell}(x_\ell)$$

for all $x_1, \ldots, x_k$.

2.7 Moments

The probability density function of a continuous or probability mass function of a discrete random variable X provides us the probabilities of all the possible values of X. It is often desirable to summarize this information in a single representative number.

To do so, we can look at the average value of X if we were to sample it many times. This value (which we call the *expectation* of X and define below) requires that the variable does not take extremely large values too often; otherwise, this average may explode and thus be ill defined. We formalize the property of a variable being nonextreme as *integrability*.

2.7.1 Nonnegative random variables

To compute the mean under a probability distribution P, we need to integrate a random variable against P. We first properly define integrals of functions taking nonnegative values. The case of nonpositive functions is identical. Then, for a general function, we will naturally obtain an integrability condition to define its integral as the difference of its positive and negative parts.

We have seen that a real random variable X is a function from a probability space $(\Omega, \mathcal{F}, P)$ to $(\mathbb{R}, \mathcal{G})$ (where, often, $\mathcal{G}$ is chosen as the Borel σ-field on $\mathbb{R}$). We have seen that this defines a PDF p_X for discrete random variables or a density for continuous random variables on $\mathbb{R}$, which characterizes the law (or distribution) of X. Hence, using p_X, we can directly compute integrals on $\mathbb{R}$ without explicitly defining $(\Omega, \mathcal{F}, P)$, e. g., to compute the average value of X.

Definition 2.7.1. Suppose X is a nonnegative random variable, that is, $P(X \geq 0) = 1$. The *expectation* of X is defined as

$$\mathbb{E}[X] = \sum_{x} x\, p_X(x) \tag{2.31}$$

if X is discrete and as

$$\mathbb{E}[X] = \int_{-\infty}^{\infty} x f_X(x) dx \tag{2.32}$$

if X is continuous.

Remark 2.7.2. The case of random variables that are neither discrete nor continuous is out of the scope of this class.

Remark 2.7.3. Note that $\mathbb{E}[X]$ does not have to be finite, in which case $\mathbb{E}[X] = +\infty$, so that $\mathbb{E}[X]$ is always well defined (as long as X is nonnegative).

Instead of integrating X, we can integrate $g(X)$, where $g : \mathbb{R} \to \mathbb{R}_+$. The only restriction is that $g(X)$ is a random variable; this depends on g and is the case for all the maps we will consider in this book. (For example, it is true as soon as g is piecewise continuous.) Hence, for **any** real random variable X and g such that $g(X)$ is a random variable, the following is well defined:

$$\mathbb{E}[g(X)] = \sum_{x} g(x) p_X(x)$$

if X is discrete, and

$$\mathbb{E}[g(X)] = \int_{-\infty}^{\infty} g(x) f_X(x) dx$$

if X is continuous.

Remark 2.7.4. The expectation $\mathbb{E}[g(X)]$ can equivalently be written on the canonical probability space $(\Omega, \mathcal{F}, P)$ as follows:

$$\mathbb{E}[g(X)] = \int_{\Omega} g(X(\omega)) P(d\omega).$$

2.7.2 The case of signed random variables

In the previous section, we introduced the expectation of a nonnegative random variable and assigned it a well-defined value. Identical arguments can be made to define the expectation of nonpositive variables. In this section, we aim at defining $\mathbb{E}[X]$ for a random variable X that can take signed values, that is, $P[X > 0] > 0$ and $P[X < 0] > 0$.

If we try to define the expectation as in the previous section, a problem can arise with infinite values. Let us see an example. Let X be such that $p_X(k) = p_X(-k) = \frac{3}{\pi^2} \cdot \frac{1}{k^2}$ for all $k \in \mathbb{N}$ and $p_X(x) = 0$ for all $x \notin \mathbb{N} \cup -\mathbb{N}$. We can check that $\sum_{k \in \mathbb{N} \cup -\mathbb{N}} p_X(k) = 1$, so that p_X is indeed a probability measure. However, if we try to define $\mathbb{E}[X]$ as $\sum_{k \in \mathbb{N} \cup -\mathbb{N}} k p_X(k)$, then the positive and negative parts will be

$$\frac{3}{\pi^2} \sum_{k \in \mathbb{N}} \frac{k}{k^2} = +\infty \tag{2.33}$$

and

$$\frac{3}{\pi^2} \sum_{k \in -\mathbb{N}} \frac{k}{k^2} = -\infty. \tag{2.34}$$

Since $+\infty - \infty$ is ill defined, we cannot make sense of $\mathbb{E}[X]$ in that case.

To avoid this issue, we need to restrict ourselves to *integrable* random variables: Let $X_+ = X\mathbf{1}_{X \geq 0}$ and $X_- = |X|\mathbf{1}_{X < 0}$, so that $X = X_+ - X_-$ (note that both X_+ and X_- are nonnegative random variables).

Definition 2.7.5. We say that a real random variable X is integrable if $\mathbb{E}[X_+], \mathbb{E}[X_-] < +\infty$. In this case, we define

$$\mathbb{E}[X] = \mathbb{E}[X_+] - \mathbb{E}[X_-]. \tag{2.35}$$

The integrability condition is often written $\mathbb{E}[|X|] < \infty$, since $\mathbb{E}[|X|] = \mathbb{E}[X_+] + \mathbb{E}[X_-]$ by the linearity of expectation (which follows from the linearity of summation/integral).

Remark 2.7.6. When exactly one of $\mathbb{E}[X_+]$ and $\mathbb{E}[X_-]$ is infinite, then it is possible to define $\mathbb{E}[X] = +\infty$ or $-\infty$ according to whether X_+ or X_- is not integrable. In this case, even though $\mathbb{E}[X]$ is well defined (infinite), X is not integrable.

We now list the basic most important properties for expectations, where X and Y are integrable:
- If $X \geq 0$, then $\mathbb{E}[X] \geq 0$ (monotonicity).
- $|\mathbb{E}[X]| \leq \mathbb{E}[|X|]$ (triangle inequality).
- If X and Y are integrable, then $\mathbb{E}[aX + bY] = a\mathbb{E}[X] + b\mathbb{E}[Y]$ for all $a, b \in \mathbb{R}$ (linearity).
- If $X = c$ then $\mathbb{E}[X] = c$.
- For any event $A \in \mathcal{F}$, we have $P(A) = \mathbb{E}[\mathbf{1}_A]$.
- If X and Y are square integrable, then $\mathbb{E}[|XY|] \leq \sqrt{\mathbb{E}[X^2]\mathbb{E}[Y^2]}$, with equality if and only if $X = aY$ for some $a \in \mathbb{R}$ (Cauchy–Schwarz inequality).

For a random variable X with $\mathbb{E}[X^2] < \infty$ (that is, X is square integrable), we can define its variance as

$$\mathrm{Var}(X) = \mathbb{E}[(X - \mathbb{E}[X])^2] = \mathbb{E}[X^2] - \mathbb{E}[X]^2.$$

- The square root of the variance is the *standard deviation*, often denoted by σ_X or just σ.
- $\text{Var}(aX) = a^2\,\text{Var}(X)$.
- If X and Y are independent and square integrable, then $\mathbb{E}[XY] = \mathbb{E}[X]\mathbb{E}[Y]$, and $\text{Var}(X + Y) = \text{Var}(X) + \text{Var}(Y)$.

The *covariance* of two square-integrable variables X and Y is defined as

$$\text{Cov}(X, Y) = \mathbb{E}[(X - \mathbb{E}[X])(Y - \mathbb{E}[Y])] = \mathbb{E}[XY] - \mathbb{E}[X]\mathbb{E}[Y]. \tag{2.36}$$

If $\text{Cov}(X, Y) = 0$, then we say that X and Y are *uncorrelated*.

Independence implies uncorrelated, but uncorrelated does not imply independence. By rescaling the covariance we obtain the *correlation* (assuming that none of X and Y is deterministic):

$$\rho(X, Y) = \frac{\text{Cov}(X, Y)}{\sqrt{\text{Var}(X)\,\text{Var}(Y)}} \in [-1, 1]. \tag{2.37}$$

The correlation $\rho(X, Y)$ can be thought of as a measure of the linear dependence between X and Y; see Figure 2.2. However, it is only informative about linear trends and does not capture more complex relationships between X and Y; see Figure 2.3.

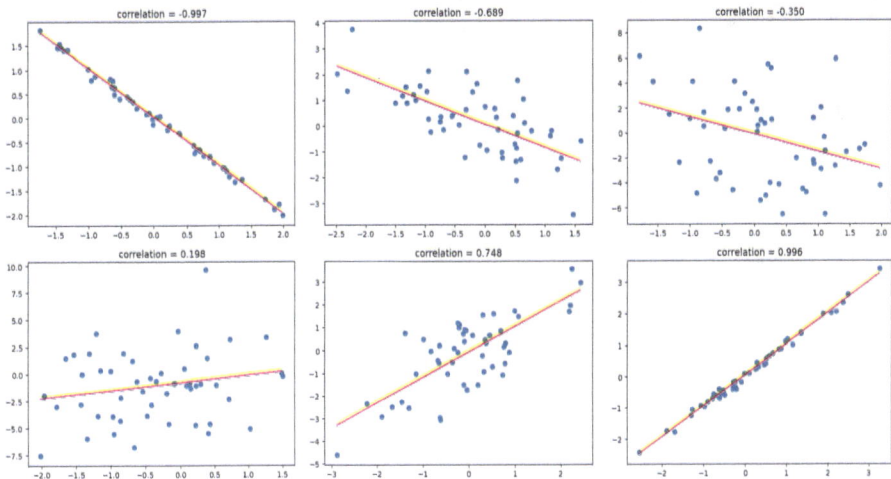

Figure 2.2: Plots of correlations for different clouds of points. The correlation **is not** the slope of the best linear fit, but the amount of noise around the best linear fit and its sign characterizes the direction of the slope (increasing or decreasing).

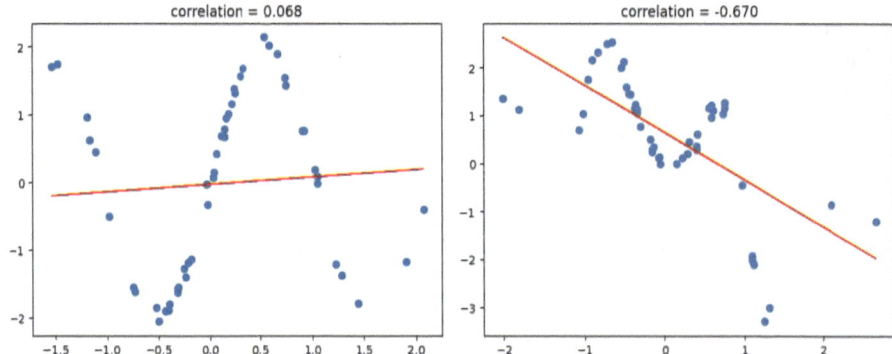

Figure 2.3: Plots of correlations for different clouds of points. The correlation fails to capture nonlinear relationship between the two variables.

Remark 2.7.7. The fact that the correlation always belongs to $[-1, 1]$ is proven using the Cauchy–Schwarz inequality. In particular, the equality case of this inequality entails that $\rho(X, Y) = 1$ (resp., -1) if and only if $Y = aX$ for some real $a > 0$ (resp., $a < 0$).

In general, with two discrete integrable random variables, we can form the joint expectation

$$\mathbb{E}[g(X, Y)] = \sum_X \sum_Y g(x, y)\, p_{X,Y}(x, y). \tag{2.38}$$

The *conditional expectation* of X given Y is defined as

$$\mathbb{E}[X|Y = y] = \sum_X x\, p_{X|Y}(x|y). \tag{2.39}$$

The *total expectation theorem* can be derived as well:

$$\sum_y \mathbb{E}[X|Y = y]\, p_Y(y) = \mathbb{E}[X]. \tag{2.40}$$

For the case of two continuous random variables, we have the joint expectation

$$\mathbb{E}[g(X, Y)] = \int_{-\infty}^{\infty} \int_{-\infty}^{\infty} g(x, y) f_{X,Y}(x, y)\, dx\, dy. \tag{2.41}$$

The *conditional expectation* of X given Y is defined as

$$\mathbb{E}[X|Y = y] = \int_{-\infty}^{\infty} x\, f_{X|Y}(x|y)\, dx. \tag{2.42}$$

The case where X is discrete and Y is continuous is similar with the integral over the values of X replaced by a sum.

For a real $p > 1$, if X^p is integrable, then its pth moment is defined as

$$\mathbb{E}[X^p]. \tag{2.43}$$

For any $p > q \geq 1$, if X^p is integrable, then X^q is also integrable. Hence the values of $p \geq 1$ such that X^p is integrable form an interval (possibly, empty).

Remark 2.7.8. The last statement about the integrability of X^q implied by that of X^p is a consequence of the following more general fact: Let $L^p = L^p(\Omega, \mathcal{F}, m)$ be the space of functions $f : \Omega \to \mathbb{R}$ such that $\int_\Omega |f(\omega)|^p m(d\omega)$ for some measure m. If m is finite (i. e., $m(\Omega) < \infty$), then $L^q \subset L^p$.

Informally, for an integral to be ill defined, we need either the integrand $f(\omega)$ to explode or the measure m to assign an infinite mass on a region where the function is not close to zero. When m is bounded, only the large values of $f(\omega)$ can be problematic. Therefore, if the values of $|f(\omega)|^p$ are not too large, then it is also the case of $|f(\omega)|^q$ since $x^q < x^p$ for $x > 1$.

Since we work with a *probability space* ($m = P$), we can just apply this fact and obtain the above-mentioned property of the moments of a random variable.

2.8 Samples from an unknown distribution

2.8.1 Law of large numbers and central limit theorem

Whereas probability theory addresses general questions related to $(\Omega, \mathcal{F}, P)$, the field of statistics is concerned with questions where the distribution of a random variable p_X is unknown. Intuitively, the more observations from p_X, the more we know about it. This intuition is formalized in the *Law of Large Numbers*:

Theorem 2.8.1. *Let $(X_i)_{i \geq 1}$ be a sequence of i. i. d. and integrable random variables. For all $m \in \mathbb{N}$, define $S_m := \sum_{i=1}^{m} X_i$. Then*

$$\lim_{m \to \infty} \frac{S_m}{m} = \mathbb{E}[X_1]$$

with probability 1.

When a statement holds with probability 1 under P, we say that the statement holds *P-almost surely*. Theorem 2.8.1 tells us is that the empirical mean of i. i. d. and integrable random variables converges to the true mean (i. e., the expectation). In other words, if we do not know p_X but have access to an infinite number of independent observations, then we can retrieve the expectation of p_X.

In practice, of course, we do not have access to infinitely many observations. The law of large numbers does not tell us at what speed the empirical mean converges. It turns out that by adding the assumption of finite second moment the *Central Limit Theorem* gives us the order of magnitude of the distance between the empirical and true means.

Theorem 2.8.2. *Let $(X_i)_{i\geq 1}$ be a sequence of i. i. d. square-integrable random variables. For all $m \in \mathbb{N}$, define $S_m = \sum_{i=1}^{m} X_i$. For any real numbers $a < b$, we have*

$$\lim_{m\to\infty} P\left[\sqrt{\frac{m}{\mathrm{Var}[X_1]}} \left(\frac{S_m}{m} - \mathbb{E}[X_1] \right) \in (a, b) \right] = P[a < Y < b],$$

where $Y \sim \mathcal{N}(0, 1)$.

In the central limit theorem, we have the term $\frac{S_m}{m} - \mathbb{E}[X_1]$, which converges almost surely to 0 as $m \to \infty$ by the law of large numbers. Multiplying this difference by the factor $\sqrt{\frac{m}{\mathrm{Var}[X_1]}}$, we get a random number with normal distribution with mean 0 and variance 1. This factor grows at the speed $\sqrt{m}$ (while $\mathrm{Var}[X_1]$ remains constant in m). This means that in the law of large numbers, the random variable $\frac{S_m}{m}$ converges to $\mathbb{E}[X_1]$ **exactly** at speed $\frac{1}{\sqrt{m}}$. The constant factor simply normalizes the Gaussian limit to have variance 1.

2.8.2 Estimating an unknown probability

In this section, we see on a concrete example how the theorems from the previous section can be used to estimate an unknown probability from a finite sample.

The frequentist way

Suppose that we have a fair coin. The fact that we specify the coin is fair implicitly poses the probability P such that $P(\text{“heads”}) = P(\text{“tails”}) = 1/2$. If instead we are given a coin and we do not know whether it is fair, then we can only say that there exists $p \in [0, 1]$ such that

$$P(\text{“heads”}) = 1 - P(\text{“tails”}) = p.$$

We can, however, say more about the likely values of p by *estimating* it through repeated experiments as follows.

We can see the outcome of a coin toss as that of a Bernoulli random variable with parameter p, where the variable is equal to 1 if the coin yields "heads" and 0 if "tails". Suppose that we toss the coin $m \in \mathbb{N}$ times and let X_i be the outcome of the ith toss. Note that $(X_i)_{i=1}^{m}$ are i. i. d. Using the law of large numbers, we know that $S_m = \sum_{i=1}^{m} X_i$ is such that $\frac{S_m}{m} \to p$ as $m \to \infty$. Hence, for a large enough m, we can estimate p by

$$\hat{p} := \frac{\#\text{ of heads}}{\#\text{ of tosses}} = \frac{S_m}{m}.$$

Naturally, if $m = 3$, then it is unlikely that our estimate is satisfactory. Suppose that we want to have an error of order $1/100$. Then we can use the central limit theorem that tells us that the error of the estimate, that is, $\hat{p} - p$, is of order $1/\sqrt{m}$ up to a random number that looks like a standard centered Gaussian for large enough m. Hence, to obtain an error of order $1/100$, we need

$$\frac{1}{\sqrt{m}} \leq \frac{1}{100} \quad \Leftrightarrow \quad m \geq 100^2 = 10000,$$

that is, by tossing ten thousand times the coin, we have that $\hat{p} - p \approx \frac{1}{100}Y$, where $Y \sim \mathcal{N}(0, 1)$.

Take-home message: In this class, "learning" means approximating a function or a distribution given a dataset. From the above example the law of large numbers and the central limit theorems may be seen as first results on learning. Assuming that the data are indeed i. i. d., the more data, the better.

The Bayesian way

The example of the coin toss above is the *frequentist* paradigm of statistics. It typically requires a large amount of data to be accurate. The other paradigm is *Bayesian* statistics, which yields effective estimate with few data but often requires more computations and an a priori distribution. It is based on Bayes' theorem.

Theorem 2.8.3 (Bayes' theorem). *Let $A, B \in \mathcal{F}$ be such that $P(A), P(B) > 0$. Then*

$$P(B|A) = \frac{P(B)P(A|B)}{P(A)}. \tag{2.44}$$

The informal interpretation of Bayes' theorem in the context of learning is the following: suppose B represents your a priori beliefs of the world and A is some observations linked to these beliefs. Ideally, you want to update your beliefs according to A. That is exactly what this theorem tells us: the probability of our beliefs B *a posteriori* (that is, after having observed A) is given by the right-hand side of the equation of the statement. Note that it depends on three terms: the prior probability we attribute to our beliefs $P(B)$, the probability of the observations given our prior beliefs $P(A|B)$, and the last term more difficult to interpret $P(A)$. This last term can further be decomposed as

$$P(A) = P(A|B)P(B) + P(A|B^c)P(B^c),$$

so that we can compute $P(A)$ according to whether our beliefs are true or not and the prior probability we assign to our beliefs.

Let us see how this works with an example.

Example 2.8.4 (Naive Bayes classifier). This is a simple "probabilistic classifier" based on applying Bayes' theorem with strong (naive) independence assumptions between the features. Let us consider again a binary classification. The naive Bayes classifier is a conditional probability model: given an input x to be classified, represented by a vector $x = (x_1, \ldots, x_m)$ representing m features, it assigns the conditional probabilities

$$p(y = +1|x_1, \ldots, x_m), \quad p(y = -1|x_1, \ldots, x_m).$$

Using Bayes' theorem, we can write

$$p(y = +1|x_1, \ldots, x_m) = \frac{p(y = +1)p(x|y = +1)}{p(x)}.$$

The numerator is equivalent to the joint probability

$$p(y = +1, x_1, \ldots, x_m),$$

which can be rewritten as follows using the chain rule for repeated applications of the definition of conditional probability:

$$
\begin{aligned}
p(y = +1, x_1, \ldots, x_m) &= p(x_1|x_2, \ldots, x_m, y = +1)p(x_2, \ldots, x_m, y = +1) \\
&= p(x_1|x_2, \ldots, x_m, y = +1)p(x_2|x_3, \ldots, x_m, y = +1) \\
&\quad \times p(x_3, \ldots, x_m, y = +1) \\
&= \cdots \\
&= p(x_1|x_2, \ldots, x_m, y = +1)p(x_2|x_3, \ldots, x_m, y = +1) \cdots \\
&\quad \times p(x_{m-1}|x_m, y = +1)p(x_m|y = +1)p(y = +1).
\end{aligned}
$$

Suppose now that the so-called **naive conditional independence assumption** holds, which tells us that all features in x are mutually independent, conditional on the label (e. g., $y = +1$ or $y = -1$). Under the assumption

$$p(x_i|x_1, \ldots, x_{i-1}, x_{i+1}, \ldots, x_m, y = +1) = p(x_i|y = +1),$$

the original probability can be rewritten as

$$
\begin{aligned}
p(y = +1|x_1, \ldots, x_m) &\propto p(y = +1, x_1, \ldots, x_m) \\
&\propto p(y = +1)p(x_1|y = +1) \cdots p(x_m|y = +1).
\end{aligned}
$$

For a new example x, we can compute our best guess as the true label using

$$\hat{y} = \arg\max_c p(y = c|x).$$

This is called the MAP estimate (maximum a posteriori).

Example 2.8.5 (Application of naive Bayes classifier). We now use the naive Bayes classifier to answer the following question: Suppose that there is an infectious disease for which we have a test with false positive probability 0.01 and false negative probability 0.2. This means that a test on a noninfected patient returns a positive result ("infected") with probability 0.01 and a test on an infected patient returns a negative result with probability 0.2. Suppose, moreover, that we know that 10 % of the population is infected.

More formally, let $y \in \{-1, 1\}$ be such that $y = 1$ if the patient is infected and -1 if they are not infected. Suppose that the patient was tested three times (let us ignore the timing of the tests for this exercise) and the results were $x_1 = 1$, $x_2 = -1$, $x_3 = 1$, where 1 means that the test was positive and -1 that it was negative.

The naive conditional independence assumption holds in this case, since given y, the results of the tests x_1, x_2, x_3 are independent (we ignore the influence of timing). We can therefore compute

$$p(y = +1|x_1 = 1, x_2 = -1, x_3 = 1)$$
$$\propto p(y = +1)p(x_1 = 1|y = +1)p(x_2 = -1|y = +1)p(x_3 = 1|y = +1)$$
$$\propto 0.1 \times (1 - 0.2) \times 0.2 \times (1 - 0.2) = 0.0128.$$

Similar computations yield $p(y = -1|x_1 = 1, x_2 = -1, x_3 = 1) \propto 0.0000891$. In particular, we deduce that $p(y = +1|x_1 = 1, x_2 = -1, x_3 = 1) > p(y = -1|x_1 = 1, x_2 = -1, x_3 = 1)$, and the naive Bayes classifier classifies the tested individual as "infected".

3 Optimization

Recall equation (1.3) in Chapter 1,

$$\min_{w \in \mathbb{R}^P} \frac{1}{m} \sum_{i=1}^{m} \ell(f_w(x_i), y_i).$$

The model f_w is parametric, and the problem consists of searching for the best parameters w within the parameter space, i. e., the parameters that minimize the loss on the dataset. This is done through optimization techniques.

This chapter is an introduction to optimization, oriented toward its application to machine learning. Some questions of interest are as follows:
-   What algorithms can be used to solve an optimization problem?
-   Is convergence guaranteed?
-   How long does it take?
-   If it converges, then is the limit always optimal?

In this book – a fortiori in this chapter – we will be concerned with local algorithms only, based on the *gradient* of the cost function evaluating the performance of the model.

3.1 Definitions

Let $C : \mathbb{R}^P \to \mathbb{R}_+$ be a generic cost function on the parameter space. The content of this chapter applies to an empirical loss such that $C(w) = L(f_w)$ evaluates the performance of a parametric function f_w on a given dataset.

3.1.1 Multidimensional derivatives

In one dimension, we can study the variation of a differentiable map $g : \mathbb{R} \to \mathbb{R}$ in the vicinity of a point x_0 through the sign of its derivative $g'(x_0)$. The same treatment can be made in the multidimensional case.

Definition 3.1.1. We say that C is *differentiable* if for all $w \in \mathbb{R}^P$ and all $i = 1, \ldots, P$, the limit $\lim_{h \to 0} C(w_1, \ldots, w_i + h, \ldots, w_P) =: \partial_{w_i} C(w)$ exists and is unique. The *gradient* of C at $\overline{w}$ is the column vector

$$\nabla_w C(\overline{w}) := \left(\partial_{w_i} C(\overline{w}) \right)_{i=1,\ldots,P}.$$

When $w \mapsto \nabla_w C(w)$ exists and is continuous, we say that C is in $C^1(\mathbb{R}^P, \mathbb{R}_+)$ or simply C^1. Henceforth, we always assume that $C \in C^1$.

https://doi.org/10.1515/9783111288994-003

In the multidimensional case, variations can be defined in multiple directions; nonetheless, the gradient is a vector whose direction has *maximal increase* in the following sense: let $w \in \mathbb{R}^P$, and let $v \in \mathbb{R}^P$ be an arbitrary vector such that $\|v\| = \|\nabla_w C(w)\|$ for the Euclidean norm. Then, for all $\epsilon > 0$ small enough, we have

$$C(w + \epsilon \nabla_w C(w)) \geq C(w + \epsilon v). \tag{3.1}$$

We see from the definition of the gradient that $-\nabla_w C(w_0)$ satisfies the converse inequality, that is, it is the direction of maximal decrease around w_0.

The gradient is the first-order term in the Taylor expansion of a smooth function in the sense that

$$C(w') - C(w) = \nabla_w C(w) \cdot (w' - w) + o(\|w' - w\|),$$

where $\cdot$ denotes the dot product in $\mathbb{R}^P$. This formula says in particular that the steepness of the slope from w in any arbitrary direction $u \in \mathbb{R}^P$ with $\|u\| = 1$ is obtained as $\nabla_w C(w) \cdot u$. Sometimes, however, the first order alone is not informative enough about the neighborhood of a point. We can think, for example, of the one-dimensional maps $x \mapsto 0, x \mapsto x^2$, and $x \mapsto x^3$, all of which have the null derivative at 0 but exhibit very different behavior around 0. More generally, $\nabla_w C(w) = (0, \ldots, 0)^{\mathsf{T}}$ does not tell us anything about the existence of strictly increasing, decreasing, or constant directions around w.

In the above one-dimensional examples, we can compute the second-order derivatives to lift the ambiguity. It works identically in higher dimensions, provided that the function is smooth enough. Henceforth, we assume that $C \in \mathcal{C}^2$, that is, C is twice differentiable with continuous second-order derivatives.

Definition 3.1.2. Suppose that $C \in \mathcal{C}^2$. The *Hessian matrix* of C at a point $w \in \mathbb{R}^P$, denoted by $\mathcal{H}_w$, is defined by

$$\mathcal{H}_w := (\partial_{w_i} \partial_{w_j} C(w))_{1 \leq i, j \leq P}.$$

A well-known result, known as Schwarz's theorem, states that if $C \in \mathcal{C}^2$, then partial derivatives can be swapped, and therefore $\mathcal{H}_w$ is symmetric.

The second-order Taylor expansion reads as

$$C(w') - C(w) = \nabla_w C(w) \cdot (w' - w) + \frac{1}{2}(w' - w)^{\mathsf{T}} \mathcal{H}_w (w' - w) + o(\|w' - w\|^2).$$

Hence, compared to (3.1), we gained precision on the approximation of $C(w')$ when w' is close to w.

How does the Hessian inform us about the variations of C? As a symmetric matrix, the Hessian only has real eigenvalues, say $\lambda_1 \geq \cdots \geq \lambda_P$. The associated eigenvectors $e_1, \ldots, e_P$ are defined by the property $\mathcal{H}_{w_0} e_i = \lambda_i e_i$. Moreover, they are orthogonal, that is, $\sum_{k=1}^{P} e_i(k) e_j(k) = 0$ for all $i \neq j$, and therefore they form a basis of $\mathbb{R}^P$. Eigenvalues and eigenvectors have a natural interpretation: moving away from w in the direction e_i induces a change in the gradient such that $\nabla_w C(w + \epsilon e_i) \approx \nabla_w C(w) + \epsilon \lambda_i e_i$. In particular, depending on the eigenvalues, it sometimes allows us to lift the ambiguity when $\nabla_w C(w_0) = 0$. This is discussed in greater detail later on in this chapter.

3.1.2 Critical points and their classification

Thanks to the previous section, we can look at the variation of C around a point w in the parameter space thanks to the gradient and the Hessian of C at w. In particular, even when the gradient vanishes at a point, the Hessian of C at the problematic point may provide enough information to locally characterize the variations of C. We now formally define a classification of points that will be useful to discuss properties of optimization algorithms later on.

Definition 3.1.3. We say that $w_* \in \mathbb{R}^P$ is a *local minimum* if there exists $\epsilon > 0$ such that $C(w_*) \leq C(w)$ for all $w \in \mathbb{R}^P$ such that $\|w_* - w\| < \epsilon$.

If $C(w_*) \leq C(w)$ for all $w \in \mathbb{R}^P$, then we say that w_* is a *global minimum*.

Reversing the inequality gives the definition of a local maximum and a global maximum. When the inequality is strict, the optimum is said to be *strict*.

Definition 3.1.4. We say that $w_* \in \mathbb{R}^P$ is a *critical point* of C if $\nabla_w C(w_*) = (0, \ldots, 0)^{\mathsf{T}}$.

Lemma 3.1.5. *For a differentiable C, if w_* is a local optimum, then $\nabla_w C(w) = (0, \ldots, 0)^{\mathsf{T}}$.*

The converse is not true: think of the one-dimensional map $x \mapsto x^3$.

Definition 3.1.6. A critical point that is not a local optimum is called a saddle point.

A critical point is either a local minimum, a local maximum, or a saddle point. In many cases the Hessian is enough to discriminate them. When all eigenvalues of $\mathcal{H}_w$ are nonnegative, we say that $\mathcal{H}_w$ is positive semidefinite. When they are strictly positive, we say that $\mathcal{H}_w$ is positive definite. The definition of negative definiteness is similar with negative eigenvalues.

Theorem 3.1.7. *Let $w_* \in \mathbb{R}^P$ be a critical point of $C \in \mathcal{C}^2$. We have:*
i) *If $\mathcal{H}_{w_*}$ is positive definite, then w_* is a strict local minimum.*
ii) *If $\mathcal{H}_{w_*}$ is negative definite, then w_* is a strict local maximum.*
iii) *If $\mathcal{H}_{w_*}$ has positive and negative eigenvalues, then w_* is a saddle point.*

Remark 3.1.8. When some eigenvalues are null, the Hessian does not allow us to tell whether a point is a local optimum or a saddle point, and we need to go to higher order (if C can be differentiated more times). The Hessian's eigenvalues dictate at which speed the gradient moves from 0 around a critical point, and therefore it is often important to distinguish between *nonstrict saddle points* and *strict saddle points*: a strict saddle point is one such that the Hessian has nonzero eigenvalues of both signs, and a nonstrict saddle allows for null eigenvalues.

3.1.3 Convexity

Recall that our goal is to find w_* such that $\min_{w \in \mathbb{R}^P} C(w) = C(w_*)$. Suppose that we have found a critical point $\overline{w}$. Firstly, we have no guarantee that it is a local minimum. Secondly, even if it were one, we generally would have no guarantee that $C(\overline{w})$ is somewhat close to $\min_{w \in \mathbb{R}^P} C(w)$; see, e. g., Figure 3.1. To avoid these considerations, we can assume nicely behaved C.

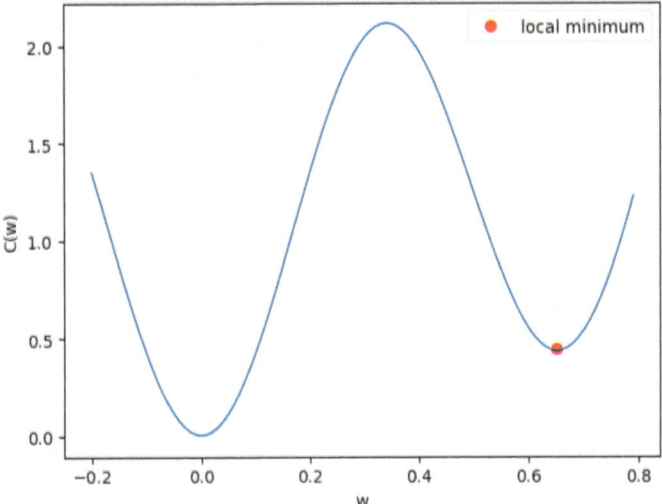

Figure 3.1: Local minimum with poor performance.

Definition 3.1.9. We say that C is *convex* (in $\mathbb{R}^P$) if for all $w, w' \in \mathbb{R}^P$, we have

$$tC(w') + (1 - t)C(w) \geq C(tw' + (1 - t)w) \quad \forall t \in (0,1).$$

If the inequality is strict, then we say that C is *strictly convex*.

Remark 3.1.10. When C is differentiable, convexity can be equivalently defined by

$$C(w') - C(w) \geq (w' - w) \cdot \nabla_w C(w) \quad \forall w, w' \in \mathbb{R}^P.$$

Geometrically, this means that the graph of C lies above its tangents; see Figure 3.2.

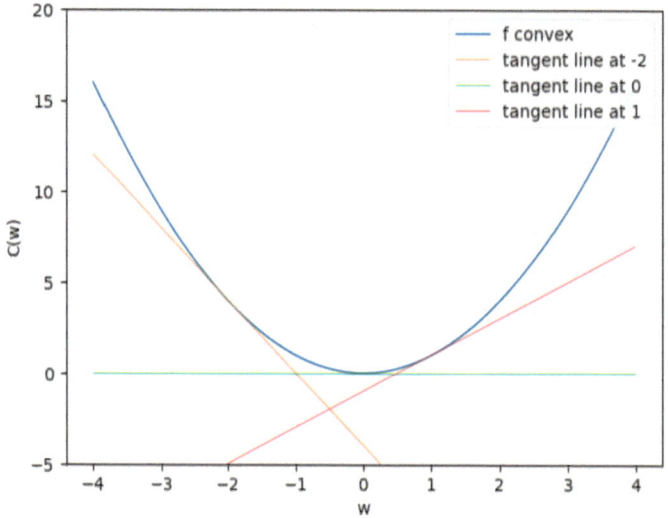

Figure 3.2: A convex map from $\mathbb{R}$ to $\mathbb{R}$ above all its tangents. This extends to multidimensional convex maps.

Proposition 3.1.11. *Suppose that $C \in C^2$ and that $\mathcal{H}_w$ is positive definite for all $w \in \mathbb{R}^P$. Then C is strictly convex. If the smallest eigenvalue $\lambda_P(w)$ of $\mathcal{H}_w$ is such that $\inf_{w \in \mathbb{R}^P} \lambda_P(w) \geq c > 0$, then we say that C is c-strongly convex, and it satisfies*

$$C(w') - C(w) \geq (w' - w) \cdot \nabla_w C(w) + \frac{c}{2}\|w' - w\|^2 \quad \forall w, w' \in \mathbb{R}^P.$$

The reader may prove Proposition 3.1.11 using the Taylor expansion of C to the second order.

Example 3.1.12. All polynomials of degree 2 with leading coefficient m from $\mathbb{R}$ to $\mathbb{R}$ are m-strongly convex. The map $x \mapsto e^x$ is strictly convex but not strongly convex. The map $x \mapsto x^4$ is strictly convex but not strongly convex. The map $x \mapsto x^3$ is not convex.

Note that the map in Figure 3.1 is nonconvex. This is not a coincidence.

Theorem 3.1.13. *If C is convex, then all critical points are global minima. If C is strictly convex, then the global minimum is unique.*

A typical example of a convex map with infinitely many global minima is $f : \mathbb{R}^2 \to \mathbb{R}$ defined by $f(x, y) = (x + y)^2$, which has a line of global minima $\{(x, y) \in \mathbb{R}^2 : x = -y\}$.

Consider a c-strongly convex map $f : \mathbb{R} \to \mathbb{R}$ achieving its global minimum at x_*. This means that $f'(x) \geq f'(x_*) + (x - x_*)f''(x_*) \geq 0 + (x - x_*)c$. Now $f(x) - f(x_*) \leq (x - x_*)f'(x) \leq \frac{1}{c}f'(x)^2$, so that $c(f(x) - f(x_*)) \leq f'(x)^2$. If f is a cost function, in particular, nonnegative, then we see that $f'(x)^2$ scales as c times the gap in performance $f(x) - f(x_*)$. This property naturally extends to multidimensional strongly convex maps.

Lemma 3.1.14. *If $C \in C^2$ is c-strongly convex for some $c > 0$, then*

$$\left\| \nabla_w C(w) \right\|^2 \geq 2c(C(w) - C(w_*)),$$

where w_* is the global minimum of C.

Remark 3.1.15. The inequality in Lemma 3.1.14 is called the *Polyak–Łojasiewicz inequality*. It bounds the squared norm of the gradient at any point by the performance gap between that point and the optimum: the poorer the performance, the greater the gradient.

Proof. Since C is c-strongly convex, by Proposition 3.1.11 we have that for all $w, w' \in \mathbb{R}^P$,

$$C(w') \geq C(w) + (w' - w) \cdot \nabla_w C(w) + \frac{c}{2} \|w' - w\|^2.$$

We minimize the right-hand side with respect to w' by setting the gradient with respect to w' to 0:

$$0 = \nabla_w C(w) + c(w' - w)$$
$$\Leftrightarrow \quad w' = w - \frac{1}{c} \nabla_w C(w),$$

which then implies that for all $w, w' \in \mathbb{R}^P$,

$$C(w') \geq C(w) - \frac{1}{c} \left\| \nabla_w C(w) \right\|^2 + \frac{1}{2c} \left\| \nabla_w C(w) \right\|^2$$
$$= C(w) - \frac{1}{2c} \left\| \nabla_w C(w) \right\|^2,$$

which concludes the proof. □

3.1.4 Lipschitz maps

Continuous maps enjoy the property that close inputs produce close outputs. However, this notion of "closeness" does not guarantee the boundedness of the variations of a continuous map. For example, the exponential map $x \mapsto e^x$ is continuous, but the difference $e^{x+\epsilon} - e^x$ tends to ∞ as $x \to \infty$, even if $x + \epsilon$ and x are close. A stronger notion of continuity is sometimes required.

Definition 3.1.16. We say that a map $f : \mathbb{R}^m \to \mathbb{R}^n$ is *L-Lipschitz continuous* for some $L > 0$ if

$$\|f(x) - f(y)\| \leq L\|x - y\| \quad \forall x, y \in \mathbb{R}^m.$$

A map whose gradient is *L*-Lipschitz is called *L-smooth*.

A map is L-Lipschitz when its variations are uniformly bounded. In particular, it can be shown that a differentiable map f is L-Lipschitz if and only if $\|\nabla f(x)\| \leq L$. A geometrical interpretation for $f : \mathbb{R} \to \mathbb{R}$ is given in Figure 3.3.

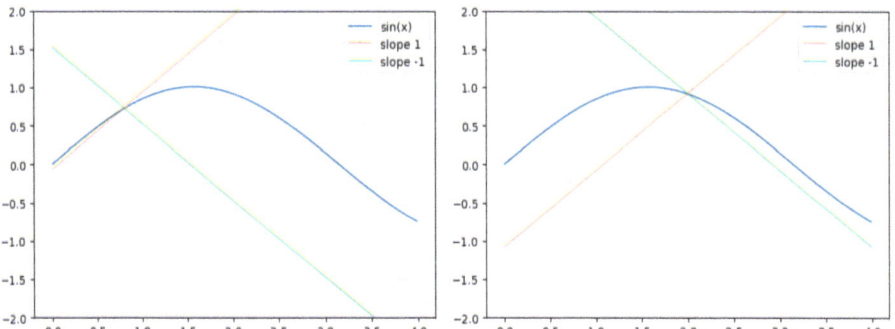

Figure 3.3: The sinus function is 1-Lipschitz: at every point $x \in \mathbb{R}$, it is contained in the cone formed by the two lines intersecting at x with slopes −1 and 1. More generally, an L-Lipschitz map is contained between such cones formed by the lines with slopes −L and L.

We saw in Proposition 3.1.11 that if the smallest eigenvalue of the Hessian of C is strictly positive, then C is strictly convex. The L-smoothness can also be seen from the Hessian:

Proposition 3.1.17. *Suppose that $C \in C^2$ and for all $w \in \mathbb{R}^P$, let $\lambda_{\max}(w)$, respectively $\lambda_{\min}(w)$, be the largest, respectively the smallest, eigenvalue of $\mathcal{H}_w$. Then*

$$\sup_{w \in \mathbb{R}^P} \{|\lambda_{\max}(w)|, |\lambda_{\min}(w)|\} \leq L < \infty \quad \Leftrightarrow \quad C \text{ is L-smooth.}$$

3.2 Gradient-based algorithms

When introducing the supervised learning framework, we informally defined the notion of training algorithms in Definition 1.2.9. Many training algorithms that are used in practice are inspired (sometimes mere variants) of the simple *gradient descent algorithm* used to solve minimization problems.

3.2.1 Gradient descent

Let $C : \mathbb{R}^P \to \mathbb{R}$ be a convex differentiable map. Recall that by Theorem 3.1.13 every critical point of C is a global minimum, i. e., $\nabla C(u) = 0$ if and only if $w_* \in \arg\min_{w \in \mathbb{R}^P} C(w)$.

The gradient descent algorithm

The gradient descent algorithm is a first-order iterative optimization algorithm for finding a local minimum of a differentiable function. Recall the notation of the ERM framework from Chapter 1: $S = \{(x_i, y_i); i = 1, \ldots, m\}$ is our dataset, and $\mathcal{H}$ is our hypothesis class.

Definition 3.2.1. Let $w_0 \in \mathbb{R}^P$ and fix $\eta > 0$. We then define w_k, $k \geq 0$, recursively as

$$w_{k+1} = w_k - \eta \nabla C(w_k).$$

One step of the above recursion is called a *gradient descent step* or *gradient descent update*. Given a number of steps $K \in \mathbb{N}$, gradient descent is the following training algorithm:

$$\mathcal{A}_{GD} : w_0 \mapsto w_K$$

for all $w_0 \in \mathbb{R}^P$, see Figure 3.4 for visualizations of gradient descent paths.

Remark 3.2.2. The positive real number η is called the *learning rate* (or *stepsize*), as it governs the size of the updates (see Figure 3.5). Both the learning rate η and the number of steps K are left unchanged when applying $\mathcal{A}_{GD}$ to w_0. Hence they are hyperparameters according to Definition 1.2.10.

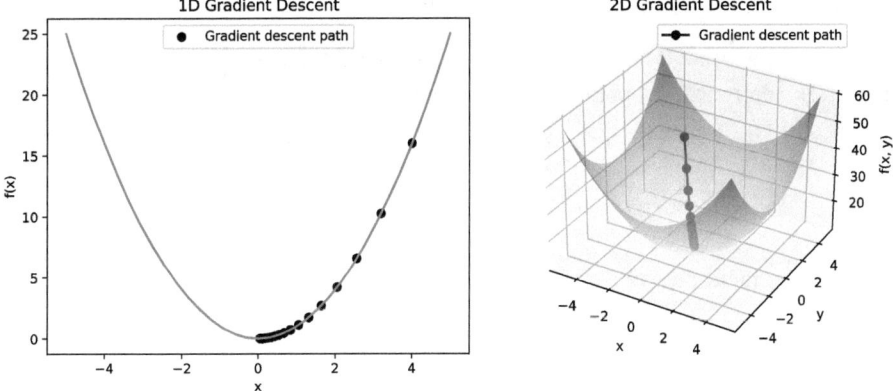

Figure 3.4: Example of gradient descent in one-dimensional problem (left) and two-dimensional problem (right).

Remark 3.2.3. In the ERM framework, the gradient descent update equation reads as

$$w_{k+1} = w_k - \eta \nabla_w L(f_{w_k}) = w_k - \eta \sum_{i=1}^{m} \nabla_w \ell(f_{w_k}(x_i), y_i).$$

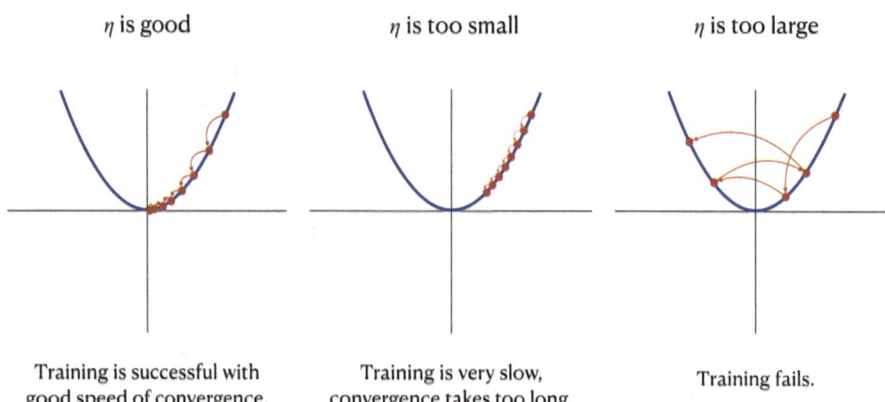

η is good η is too small η is too large

Training is successful with good speed of convergence.   Training is very slow, convergence takes too long.   Training fails.

Figure 3.5: Effect of the size of the learning rate η during training.

Gradient descent improves the performance of the model by locally moving the parameters w in the direction that maximizes the decrease of the loss. To guarantee that the gradient does not explode, it is often assumed that C is L-smooth, that is, its gradient is L-Lipschitz; see Definition 3.1.16. Some theoretical guarantees can be obtained from the L-smoothness of C.

Theorem 3.2.4. *Consider gradient descent on $C \in C^2$ with learning rate $\eta > 0$ and number of steps $K \in \mathbb{N}$. If C is L-smooth and $\eta < 2/L$, then the cost function under gradient descent converges, that is, $\lim_{K \to \infty} C(w_K)$ exists, and $\lim_{K \to \infty} \nabla_w C(w_K) = 0$.*

Proof. For this proof, we need to use the following Taylor expansion of C: for every $w, w' \in \mathbb{R}^P$, there exists a point w'' on the segment $[w, w']$ (i. e., $w'' = (1 - t)w + tw'$ for some $t \in [0, 1]$) such that

$$C(w') = C(w) + (w' - w) \cdot \nabla_w C(w) + \frac{1}{2}(w' - w)^{\mathsf{T}} \mathcal{H}_{w''}(w' - w).$$

Using the L-smoothness of C and Proposition 3.1.17, we get that

$$C(w') \leq C(w) + (w' - w) \cdot \nabla_w C(w) + \frac{L}{2}\|w' - w\|^2.$$

This translates on the gradient descent update as

$$
\begin{aligned}
C(w_{k+1}) &\leq C(w_k) + (w_{k+1} - w_k) \cdot \nabla_w C(w_k) + \frac{L}{2}\|w_{k+1} - w_k\|^2 \\
&= C(w_k) - \eta\|\nabla_w C(w_k)\|^2 + \eta^2 \frac{L}{2}\|\nabla_w C(w_k)\|^2 \\
&= C(w_k) - \eta\left(1 - \eta\frac{L}{2}\right)\|\nabla_w C(w_k)\|^2. \tag{3.2}
\end{aligned}
$$

By choosing $\eta < 2/L$ we are guaranteed to have $C(w_{k+1}) < C(w_k)$ whenever $\nabla_w C(w_k) \neq 0$. In other words, gradient descent always improves the cost when $\eta < 2/L$, and letting $K \to \infty$ yields the claim, since C is bounded from below by 0. □

The above theorem shows that under gradient descent, $C(w_K)$ converges; however, it does not say that w_K converges, as it can, for instance, diverge to ∞. (We will see in Chapter 11 that this can be the case for some Reinforcement Learning algorithm.) Note also that even if w_K converges to a critical point, it is not informative on the optimality of this critical point, nor on the rate of convergence. Convexity allows us to obtain a more detailed picture of gradient descent.

Theorem 3.2.5. *Suppose that $C \in C^2$ is L-smooth and convex and that $\eta \leq \frac{1}{L}$. Then there exists a global minimum w_* such that*

$$0 \leq C(w_K) - C(w_*) \leq \frac{\|w_* - w_0\|^2}{2\eta K}.$$

Proof. Using $\eta \leq 1/L$, equation (3.2) yields

$$C(w_{k+1}) \leq C(w_k) - \frac{\eta}{2}\|\nabla_w C(w_k)\|^2. \tag{3.3}$$

Theorem 3.2.4 entails that gradient descent converges toward a critical point w_*, and by convexity, w_* must be a global optimum. Convexity implies that

$$C(w_*) \geq C(w_k) + (w_* - w_k) \cdot \nabla_w C(w_k).$$

Combining the two bounds above entails that

$$\begin{aligned}
C(w_{k+1}) - C(w_*) &\leq (w_k - w_*) \cdot \nabla_w C(w_k) - \frac{\eta}{2}\|\nabla_w C(w_k)\|^2 \\
&\leq \frac{1}{2\eta}\left(2\eta(w_k - w_*) \cdot \nabla_w C(w_k) - \eta^2\|\nabla_w C(w_k)\|^2\right) \\
&= \frac{1}{2\eta}\left(-\|w_k - w_* - \eta\nabla_w C(w_k)\|^2 + \|w_k - w_*\|^2\right) \\
&= \frac{1}{2\eta}\left(-\|w_{k+1} - w_*\|^2 + \|w_k - w_*\|^2\right),
\end{aligned}$$

where we used that $w_{k+1} = w_k - \eta\nabla_w C(w_k)$ for the last identity. Thus the above bound shows that

$$\begin{aligned}
\sum_{k=0}^{K}(C(w_k) - C(w_*)) &\leq \frac{1}{2\eta}\sum_{k=0}^{K-1}(\|w_k - w_*\|^2 - \|w_{k+1} - w_*\|^2) \\
&= \frac{1}{2\eta}(\|w_0 - w_*\|^2 - \|w_K - w_*\|^2)
\end{aligned}$$

$$\leq \frac{\|w_0 - w_*\|^2}{2\eta}.$$

Note that $C(w_{k+1}) \leq C(w_k)$ for all $k \geq 0$, as we showed in the proof of Theorem 3.2.4. Therefore

$$C(w_K) - C(w_*) \leq \frac{\|w_0 - w_*\|^2}{2\eta K},$$

which concludes the proof. $\qquad\qquad\qquad\qquad\qquad\qquad\qquad\qquad\qquad\qquad\qquad\square$

Under the conditions of the theorem, gradient descent converges at rate $O(1/K)$. Consider the following examples from $\mathbb{R}$ to $\mathbb{R}$: $f : x \mapsto x^2$ and $g : x \mapsto x^4$ (technically, g is not L-smooth, but provided that the learning rate is small enough, gradient descent will not visit points that are too far away from 0, and on this bounded domain, g is L-smooth). The behavior of gradient descent in the vicinity of 0 is rather different for the two maps:

$$x_{k+1}^f = x_k^f - \eta 2 x_k^f = (1 - 2\eta) x_k^f,$$
$$x_{k+1}^g = x_k^g - \eta 4 (x_k^g)^3 = (1 - 4(x_k^g)^2) x_k^g.$$

We see that although the update for f remains of the order of the error $2\eta x_k^f$, that of g vanishes much quicker, that is, of the order of the cube of the error, $4\eta (x_k^g)^3$, causing a slowdown in the convergence of the algorithm. The reason can be seen from the second derivatives of f and g: both maps are convex with a unique global minimum at 0, but f is 2-strongly convex, whereas g is only convex.

The next theorem formalizes this observation.

Theorem 3.2.6. *Suppose that $C \in C^2$ is L-smooth and c-strongly convex and that $\eta \leq \frac{1}{L}$. Then*

$$0 \leq C(w_K) - C(w_*) \leq (1 - \eta c)^K (C(w_0) - C(w_*)),$$

where w_* is the unique global minimum of C.

Remark 3.2.7. Note that the assumptions on C can be translated, thanks to Propositions 3.1.11 and 3.1.17, to the assumptions that the largest and smallest eigenvalues $\lambda_1(w)$ and $\lambda_p(w)$ of its Hessian satisfy

$$\sup_{w \in \mathbb{R}^P} \lambda_1(w) \leq L, \quad \inf_{w \in \mathbb{R}^P} \lambda_p(w) \geq c.$$

Proof. Using Lemma 3.1.14 and (3.3), we get

$$C(w_{k+1}) \leq C(w_k) - \eta c (C(w_k) - C(w_*))$$
$$\Leftrightarrow \quad C(w_{k+1}) - C(w_*) \leq (1 - \eta c)(C(w_k) - C(w_*)).$$

Iterating the inequality entails that

$$C(w_K) - C(w_*) \le (1 - \eta c)^K (C(w_0) - C(w_*)),$$

which proves the claim and concludes the proof. ◻

For a more complete treatment of convergence properties of gradient descent, we refer to [8, Chapter 9].

3.2.2 Gradient flow

Closely related to gradient descent is the *gradient flow*, which can be seen as a continuous version of gradient descent when the learning rate goes to 0, as we will see. The interest of gradient flow is purely theoretical: no discussion about the learning rate is needed to prove the convergence, which then can be sometimes deduced for gradient descent with small enough learning rate.

Definition 3.2.8. We say that a family $(u(t))_{t \in \mathbb{R}_+} \subset \mathbb{R}^P$ follows the (negative of the) gradient flow of C if it is a solution of the following differential equation:

$$\partial_t u(t) = -\nabla_u C(u(t)).$$

Because $u(t)$ follows the negative of the gradient of C, the map $t \mapsto C(u(t))$ is decreasing until it reaches its limit as $t \to \infty$ (recall that C is bounded from below by 0). Assuming that $u(t) \to u_* \in \mathbb{R}^P$ as $t \to \infty$, we see that $\nabla C(u_*) = 0$, and the convexity of C ensures that $C(u_*)$ is a global minimum.

To see that gradient descent is a discrete approximation of gradient flow, let $k \in \mathbb{N}$ and use Definition 3.2.1 to write

$$w_k = -\eta \sum_{\ell=0}^{k} \nabla C(w_\ell).$$

By choosing $k = \lfloor t/\eta \rfloor$, where $\lfloor \cdot \rfloor$ denotes the integer part, we can take the limit as $\eta \to 0+$, and (assuming that ∇C is continuous to define the Riemann integral) there exists $w : \mathbb{R}_+ \to \mathbb{R}^P$ such that

$$\lim_{\eta \to 0+} w_{\lfloor t/\eta \rfloor} = \lim_{\eta \to 0+} -\eta \sum_{\ell=0}^{\lfloor t/\eta \rfloor - 1} \nabla C(w_\ell) = \int_0^t -\nabla C(w(s)) ds,$$

and therefore rescaled time gradient descent converges to gradient flow.

3.2.3 Stochastic gradient descent

In machine learning the number of trainable parameters can be large. Even when computing the gradient at an input datapoint $\nabla_w f_w(x_i)$ is relatively cheap, the size m of the dataset can dramatically increase the computational cost of gradient descent, since m gradients need to be computed at each step.

To reduce the computational cost of training, *stochastic gradient descent* samples a *single sample* of the training dataset and performs a gradient descent step on this datapoint only.

Definition 3.2.9. Let $w_0 \in \mathbb{R}^P$ and fix $\eta > 0$. Suppose that $C(w) = \frac{1}{m} \sum_{i=1}^m \ell(f_w(x_i), y_i)$. Let $(\tilde{x}_k, \tilde{y}_k)_{k \geq 0}$ be i.i.d. pairs sampled uniformly at random in the dataset $S = \{(x_i, y_i); i = 1, \ldots, m\}$. For all $k \geq 0$, define

$$w_{k+1} = w_k - \eta \nabla_w \ell(f_w(\tilde{x}_k), \tilde{y}_k).$$

Stochastic gradient descent with learning rate $\eta > 0$ and number of steps K is the training algorithm

$$\mathcal{A}_{SGD} : w_0 \mapsto w_K$$

for all $w_0 \in \mathbb{R}^P$.

Remark 3.2.10. Fix w_0 and let $(w_k^{GD})_{k \leq K}$ be the sequence generated from gradient descent, and, similarly, let $(w_k^{SGD})_{k \leq K}$ be generated by stochastic gradient descent. Stochastic gradient descent treats the dataset S as an empirical distribution, from which it samples random variables (X, Y). In particular, since we assume that S consists of i.i.d. samples from an unknown true distribution D, we can check that for all $k \geq 0$,

$$\mathbb{E}[w_{k+1}^{SGD} | w_k^{SGD} = w_k^{GD}] = w_{k+1}^{GD}.$$

In other words, the expectation of the SGD update is equal to the GD update. However, one SGD update **does not** necessarily make C decrease, that is, possibly, $\mathbb{P}[C(w_{k+1}^{SGD}) \leq C(w_k^{SGD})] < 1$.

Remark 3.2.11. Many variations of SGD can be derived from the same idea. For example, instead of sampling a single example each step, we can randomly sample a *minibatch* of examples. The previous remark remains correct, that is, the expectation of SGD is GD by using a minibatch, and the larger the batch, the smaller the variance of the SGD update.

The following convergence rates and their proofs for SGD can be found, e.g., in [22, Theorems 5.5 and 5.7].

Theorem 3.2.12. *Suppose that* $C \in \mathcal{C}^2$ *and that there exists* $w_* \in \mathbb{R}^P$ *such that* $C(w_*) = \frac{1}{m} \sum_{i=1}^m \ell(f_{w_*}(x_i), y_i) = 0$. *Let* $(w_k)_{k \geq 0}$ *be generated by SGD on* C, *and let* $\overline{w}_k := \frac{1}{k} \sum_{i=0}^{k-1} w_i$.

(i) *If $w \mapsto \ell_w(x_i)$ is L-smooth and convex for all $i = 1, \ldots, m$ and if $\eta < \frac{1}{2L}$, then for all $k \geq 1$,*

$$\mathbb{E}[C(\overline{w}_k)] \leq \frac{\|w_0 - w_*\|^2}{2\eta(1 - 2\eta L)k}.$$

(ii) *If $w \mapsto \ell_w(x_i)$ is L-smooth and c-strongly convex for all $i = 1, \ldots, m$ and if $\eta < \frac{1}{2L}$, then for all $k \geq 1$,*

$$\mathbb{E}[\|w_k - w_*\|^2] \leq (1 - \eta c)^k \|w_0 - w_*\|^2.$$

3.2.4 Momentum

To speed up the optimization algorithm, the history of the algorithm can be taken into account in the update. Many forms can be considered; we focus on the so-called *Heavy Ball Algorithm*:

$$w_{k+1} = w_k - \eta \nabla C(w_k) + \beta(w_k - w_{k-1}).$$

The last term keeps in memory the last update, scaled by a factor $\beta \in [0, 1)$, and is therefore interpreted as the momentum. When $\beta = 0$, we get GD back. To see how the momentum affects the training trajectory, we write

$$w_1 - w_0 = -\eta \nabla C(w_0),$$
$$w_2 - w_1 = -\eta \nabla C(w_1) + \beta(w_1 - w_0) = -\eta(\nabla C(w_1) + \beta C(w_0)),$$
$$\vdots$$
$$w_k - w_{k-1} = -\eta \sum_{i=1}^{k} \beta^{k-i} \nabla C(w_{i-1}).$$

We see that the kth update is a weighted sum of all past gradients, the older ones being discounted more than the newer ones. Small $\beta > 0$ forget faster the previous gradients, whereas the momentum is strong for large $\beta < 1$.

The momentum has the following immediate benefit: if w_k is a saddle point of C, then even though $\nabla C(w_k) = 0$, the update is likely to satisfy $w_{k+1} \neq w_k$. Hence it is easier to escape from saddle points while still converging to a nearby minimum, the momentum eventually vanishing since $0 \leq \beta < 1$, see Figure 3.6. On the other hand, if the directions of the recent previous gradients roughly align, then the momentum accelerates the update in a meaningful direction.

Formal convergence theorems can be proved for the heavy ball algorithm. In particular, akin to GD and SGD, it converges for L-smooth convex maps at a rate $O(1/k)$ and for L-smooth c-strongly convex maps at a rate $O(r^k)$ for some $r \in (0, 1)$; see, e. g., [23].

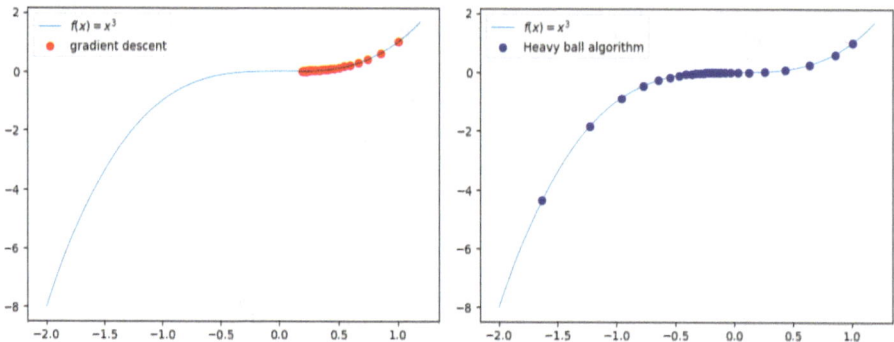

Figure 3.6: Comparison of gradient descent and heavy ball algorithms near a saddle point. For $\eta = 0.01$ and $\beta = 0.7$, both are initialized at $x = 1$ and are stopped after 27 steps. *Left*: gradient descent slows down in the vicinity of the saddle point at 0. *Right*: Heavy ball passes over the saddle point thanks to the momentum induced by the previous updates.

3.3 Nonconvex case

Recall that in the ERM framework, the cost is of the form $C(w) = L(f_w) = \frac{1}{m}\sum_{i=1}^{m} \ell(f_w(x_i), y_i)$. The mean squared error corresponds to the choice $\ell(x,y) = \|f_w(x_i) - y_i\|^2$, which is convex in both arguments. However, the parameterisation $w \mapsto f_w$ is *not necessary*, so that choosing ℓ convex does not guarantee the convexity of the cost C.

Without convexity, local minima can exist in the loss landscape. Nonetheless, several things can be studied, and in this section, we informally discuss two: the likelihood of converging to a saddle point and the rate of convergence to a critical point.

3.3.1 Strict saddle points are avoided

Consider the map $f : \mathbb{R}^2 \to \mathbb{R}$ defined by

$$f(x,y) := x^2 - y^2.$$

One can check that f is not convex and $\nabla f(0,0) = (0,0)^{\mathsf{T}}$, since $\nabla f(x,y) = (2x, -2y)^{\mathsf{T}}$. In particular, $(0,0)$ is the unique critical point, and in fact it is a strict saddle point.

Let $w_0 = (x,0)$ for any $x \neq 0$. One gradient descent step leads to

$$w_1 = w_0 - \eta \nabla f(x,0) = (x,0) - \eta(2x,0) = (x(1 - 2\eta), 0).$$

It is easy to show that $w_k \in \mathbb{R} \times \{0\}$ for all $k \geq 0$, so that for a small enough learning rate $\eta > 0$, gradient descent converges to $w_\infty = (0,0)$ (any of Theorems 3.2.4, 3.2.5, and 3.2.6 applies on the restriction of f to the line $\mathbb{R} \times \{0\}$). However, by initializing w_0 at random – say, a standard two-dimensional Gaussian – then with probability 1 the y-component of w_0 is nonnull, and we see that the y component of w_k goes away from

0 under gradient descent. Hence, with probability 1 under random initialization (that has no atom at 0, e. g., any measure absolutely continuous with respect to the Lebesgue measure), gradient descent avoids the strict saddle point at 0.

The above example is not isolated. In fact, in [31], it is proved (among other things) that

> For an L-smooth cost $C \in C^2$, the set of $w_0 \in \mathbb{R}^P$ such that $\lim_{k \to \infty} w_k$ is a strict saddle point has Lebesgue measure 0.

Several extensions of this result have been established, since, for example, we can consider vanishing learning rates; see [36] and references therein.

3.4 Constrained optimization: duality theory

In the previous sections, we have considered optimization problems without constraints on the parameters $w \in \mathbb{R}^P$. In full generality, this is not the case, and we have to optimize a certain cost function while parameters satisfy some restrictive conditions. The content of this section will be crucial when studying the *Support Vector Machine* problem in Chapter 5.

3.4.1 Lagrangian function

Consider the following constrained optimization problem:

$$\min_{w \in \mathbb{R}^P} \quad C(w) \tag{3.4}$$
$$\text{s. t.} \quad h_i(w) = 0, \; i = 1, \dots, r,$$

where the h_i are C^1 constraints (maps from $\mathbb{R}^P$ to $\mathbb{R}$).

We can use the method of *Lagrange multipliers* to solve (3.4) by turning a constrained optimization into an unconstrained optimization and introducing penalties on the violation of the constraints.

Definition 3.4.1. The *Lagrangian function* of (3.4) is defined for all $w \in \mathbb{R}^n$ and $\alpha \in \mathbb{R}^r$ by

$$\mathcal{L}(w, \alpha) = C(w) - \sum_{i=1}^{r} \alpha_i h_i(w). \tag{3.5}$$

The coefficients α_i are called the *Lagrange multipliers*.

Intuition (hand-wavy)

Suppose that w^* is the solution of (3.4). In particular, the constraints are satisfied, and for each $i \in \{1, \ldots, r\}$, if we move in a direction w orthogonal to $\nabla_w h_i(w^*)$, then we locally have $h_i(w^* + \epsilon w) \approx h_i(w^*) + O(\epsilon^2)$. Now this reasoning applies if w is orthogonal to $\nabla_w h_i(w^*)$ simultaneously for all $i \in \{1, \ldots, r\}$. If, moreover, $w^T \nabla_w C(w^*) \neq 0$, then we found a direction that decreases the cost in a neighborhood of w_* while respecting the constraints, which contradicts that w^* is optimal. This means that, necessarily, $w^T \nabla_w C(w^*) = 0$. We deduce that the gradient of the cost at w^* is orthogonal to any w orthogonal to the gradient of the constraints at w^*. In other words, $\nabla_w C(w^*) \in \mathrm{Span}(\{\nabla_w h_i(w^*); i = 1, \ldots, r\})$, that is, there exist $a_1^*, \ldots, a_r^*$ such that

$$\nabla_w C(w^*) = \sum_{i=1}^{r} a_i^* \nabla_w h_i(w^*). \tag{3.6}$$

See Figure 3.7 for a visual example.

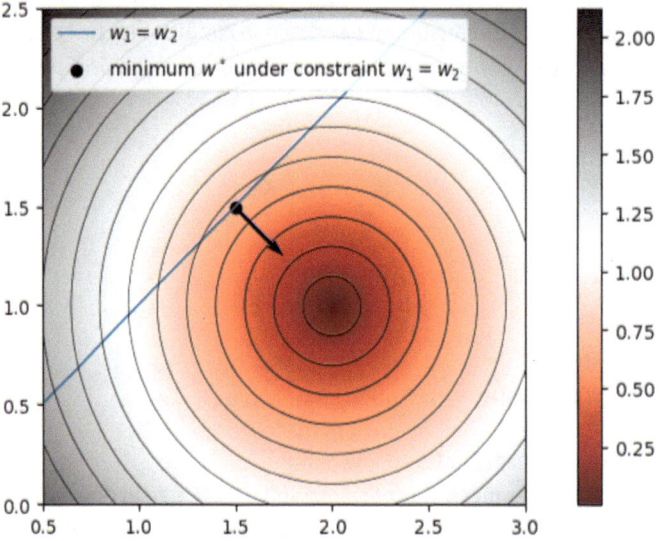

Figure 3.7: Optimizing $C(w_1, w_2) := \|(w_1, w_2) - (2,1)\|$ subject to $h(w_1, w_2) = w_1 - w_2 = 0$. The gradients $\nabla h(w^*)$ and $\nabla C(w^*)$ align at the optimum.

The above intuition justifies the definition of the Lagrangian function, as it helps us write (3.6) in a concise manner as $\nabla_w \mathcal{L}(w^*, a^*) = 0$. Computing the gradient of $\mathcal{L}(w, a)$ with respect to w and a and setting it to 0, we can solve for the unknowns. Whether we are minimizing or maximizing $\mathcal{L}$, the a_i play a role of penalizing the solution if the constraints are violated. Consider $\nabla_{w,a} \mathcal{L} = 0$ and write

$$\nabla_{w,a} C(w) - \nabla_{w,a} \sum_i a_i h_i(w) = 0.$$

Expanding this out, we have a vector where the first n entries with respect to w lead to

$$\nabla_w C(w) - \sum_i a_i \nabla_w h_i(w) = 0,$$

and the last r entries lead to

$$h_i(w) = 0, \quad i = 1, \ldots, r.$$

Define $g_{\text{prim}} : w \mapsto \max_a \mathcal{L}(w, a)$. The solution to (3.5)[1] is given in terms of the Lagrangian as

$$p^* = \min_w g_{\text{prim}}(w). \tag{3.7}$$

How does this solution relate to that of the original problem (3.4)? When considering $g_{\text{prim}}(w)$, we are fixing a w and then maximize over a. We see that as soon as one constraint is violated, say $h_i(w) \neq 0$, then we can let a_i go to plus or minus infinity (depending on the sign of $h_i(w)$) to make the Lagrangian blow up, and therefore $g_{\text{prim}}(w) = +\infty$. In particular, $g_{\text{prim}}(w)$ is finite if and only if w satisfy the constraints (provided that $C(w)$ is finite). Hence the solution w_* of (3.4) is the same as that of (3.7).

More generally, we can write the Lagrangian for constrained optimization problem that includes both equality and inequality constraints: consider

$$\min_w \quad C(w), \tag{3.8}$$
$$\text{s.t.} \quad h_i(w) = 0, \; i = 1, \ldots, r,$$
$$\text{s.t.} \quad f_i(w) \leq 0, \; i = 1, \ldots, p.$$

Definition 3.4.2. The Lagrangian function of (3.8) is defined for all $w \in \mathbb{R}^n$, $a \in \mathbb{R}^r$, and $\beta \in [0, \infty)^p$ by

$$\mathcal{L}(w, a, \beta) = L(w) + \sum_{i=1}^{r} a_i h_i(w) + \sum_{i=1}^{p} \beta_i f_i(w), \tag{3.9}$$

where the a_i and β_i are called the Lagrange multipliers.

Again, note that if any constraint is violated (i.e., if $h_i(w) > 0$ or $f_i(w) \neq 0$), then we can make the Lagrangian blow-up by sending one of $a_i \to \infty$, $\beta_i \to \infty$, or $\beta_i \to -\infty$.

Similarly as in the equality constraints, the solution to (3.8) is given in terms of the Lagrangian as

$$p^* = \min_w g_{\text{prim}}(w) = \min_w \max_{a \in \mathbb{R}^p, \beta \in [0,\infty)^r} \mathcal{L}(w, a, \beta). \tag{3.10}$$

---

1 Also called as the *primal problem/primal solution*.

The primal problem, however, does not seem easier to solve than the original problem itself. Indeed, in the minimization–maximization problem (3.10), because we maximize over α and β after having chosen a w, when minimizing over w, we are restricted to candidates that satisfy the constraints (otherwise, the Lagrangian blow-up as explained before). We would like to first choose α, β and only then minimize over w. This is called the *dual optimization problem*, which turns the constrained min-max optimization problem into a penalized max/min optimization problem where we search for an optimal solution over α, β after minimizing over w.[2]

First, we write the dual function

$$g_{\text{dual}}(\alpha, \beta) = \min_{w} \mathcal{L}(w, \alpha, \beta)$$

and then the *dual optimization problem*

$$d^* = \max_{\alpha \in \mathbb{R}^p, \beta \in [0,\infty)^r} g_{\text{dual}}(\alpha, \beta) = \max_{\alpha \in \mathbb{R}^p, \beta \in [0,\infty)^r} \min_{w} \mathcal{L}(w, \alpha, \beta). \tag{3.11}$$

How does solving this relate to solving (3.10)? We can note that the solution to (3.11) returns a lower bound to (3.10): firstly, by definition, for all w, α, β, we have

$$\mathcal{L}(w, \alpha, \beta) \leq g_{\text{prim}}(w).$$

Taking the minimum over w yields

$$g_{\text{dual}}(\alpha, \beta) \leq p_*,$$

and then the maximum over α, β gives

$$d^* \leq p^*.$$

In some cases, the solution to the original optimization problem (primal) is the same as the solution to the dual problem, i. e.,

$$d^* = p^*.$$

This is called the *strong duality condition*. To ensure strong duality, we need some conditions on the candidate solutions, namely the **Karush–Kuhn–Tucker (KKT) conditions:**[3] we say that w_* and α_* fulfil the KKT conditions if

$$\nabla_w \mathcal{L}(w^*, \alpha^*, \beta_*) = 0 \quad \text{(stationarity)},$$

---

2 If you are interested in this, then start with the notion of duality in linear programming problems.
3 Subject to differentiability and convexity requirements.

$$h_i(w^*) = 0, \quad i = 1, \ldots, r,$$
$$f_i(w_*) \le 0, \quad i = 1, \ldots, p \quad \text{(primal feasibility)},$$
$$\beta_i^* \ge 0, \quad i = 1, \ldots, p \quad \text{(dual feasibility)},$$
$$\beta_i^* f_i(w^*) = 0, \quad i = 1, \ldots, p \quad \text{(complementary slackness)}.$$

We admit the following result.

Theorem 3.4.3. *For an optimization problem for which the strong duality condition holds, any primal optimal solution w^* and dual optimal solution α^* respect the KKT conditions. Conversely, if the f_i and h_i are affine for all i, then the KKT conditions are sufficient for duality.*

The proof can be found in [8].

Part II: **Supervised learning**

4 Statistical learning theory

In this chapter, we take the point of view of statistics to study learning theory. In short, statistics is concerned with estimating an unknown distribution μ from its observations. Typical questions can be: how many data points should we collect to ensure that an estimator $\hat{\mu}$ of μ is close enough (in some sense) to μ? How does the complexity of a model relate to its empirical error and its generalization error? More generally, we are interested in deriving approximation bounds of hypothesis classes.

Remark 4.0.1. If you are familiar with numerical analysis, then you can think of this chapter as techniques to come up with *a priori error estimates* (i. e., before we sample the dataset), whereas in the previous chapter, we computed *a posteriori error estimates* (we measure the error for a specific model and a specific dataset).

4.1 Learnability

In this chapter, except where explicitly stated, we restrict ourselves to the case of binary classification, that is, $\mathcal{Y} = \{0, 1\}$, as it simplifies the presentation. As usual, we assume that the dataset $S = \{(x_i, y_i) : i \in \{1, \ldots, m\}\}$ is such that the x_i are i. i. d. with common law D in $\mathcal{X}$ and that $y_i = f(x_i)$ for some labeling function f. Furthermore, we assume the 0–1 loss function $\mathbb{1}_{\{h(x) \neq y\}}$.

Since a predictor h is trained to minimize the empirical error $L(h)$ (see (1.3)), which depends on the training set S, and since S is generated through random samplings under D, there will be randomness on the trained predictor h, and therefore in the generalization error $R_D(h)$ defined in (1.1). Thus we can see $R_D(h)$ as a random variable. Hence if we expect to find a good classifier, then we need S to be a "good representative" of D.

Example 4.1.1. Suppose we have an urn with 30 % red and 70 % green balls, and we take a sample where we get five greens "G G G G G". In this case, our sample does not represent the underlying distribution of the balls. (Note that the probability of sampling this dataset is $0.7^5 \approx 0.16$, which is far from negligible.)

From the law of large numbers (Theorem 2.8.1) we know that more data samples ensure that the dataset is representative enough to avoid situations like in the above example. In Chapter 1, we saw ways to assess the quality of a model and how to select a good model among a collection of models to solve a given task. In this chapter, we are instead concerned with studying the *learnability* of a given hypothesis class $\mathcal{H}$ from a finite dataset.

https://doi.org/10.1515/9783111288994-004

4.1.1 Realisability assumption

Definition 4.1.2 (Realisability assumption). For a given hypothesis class $\mathcal{H}$, data distribution D, and labeling function f, we say that the *realisability assumption* holds for $\mathcal{H}, D, f$ if there exists $h \in \mathcal{H}$ such that $R_{D,f}(h) = 0$.

Informally, this means that the labeling function f can be represented by elements of $\mathcal{H}$, at least on the support of D. Alternatively, if $f \in \mathcal{H}$, then clearly the realisability assumption holds.

Example 4.1.3. It is easy to construct a case where the realisability assumption does not hold: let $\mathcal{X} = \mathbb{R}$, $D = \text{Unif}(-1, 1)$, $f : x \mapsto \mathbb{1}_{\{|x|>1/2\}}$, and $\mathcal{H} = \{h_w : x \mapsto \mathbb{1}_{\{x>w\}}\}$, where $w \in \mathbb{R}$ denotes the only parameter of the model. We can check that the realisability assumption does not hold. (Note that m being fixed, there is a positive probability on the dataset S that there exists $h \in \mathcal{H}$ such that $L(h) = 0$, but this probability is not 1.)

4.1.2 PAC learnability

The realisability assumption ensures that there is a predictor in the hypothesis class that perfectly fits the data. This guarantee, however, does not mean that we are able to find this predictor. We need stronger properties for the hypothesis class. The *Probably Approximately Correct learning* (PAC learning) framework is introduced for this purpose.

For a given hypothesis class $\mathcal{H}$ and a given number of samples m, let $\mathcal{A}_{\text{ERM}}$ be any algorithm that returns a hypothesis $h_{\mathcal{A}}$ minimizing the empirical loss L as in (1.3), that is, $h_{\mathcal{A}} \in \arg\min_{h \in \mathcal{H}} L(h)$.

Definition 4.1.4 (PAC learnability). A hypothesis class $\mathcal{H}$ is PAC learnable if there exists a function $m_{\mathcal{H}} : (0, 1)^2 \to \mathbb{N}$ with the following properly: for all $\epsilon, \delta \in (0, 1)$, for every distribution $\mathcal{D}$ over $\mathcal{X}$, and for every labeling function $f : \mathcal{X} \to \{0, 1\}$, if the realisability assumption holds with respect to $\mathcal{H}, D, f$, then when running $\mathcal{A}_{\text{ERM}}$ on $m \geq m_{\mathcal{H}}(\epsilon, \delta)$ i.i.d. examples generated by D labeled by f, with probability at least $1 - \delta$, we have $R_D(h_{\mathcal{A}}) \leq \epsilon$.

We have two parameters in the PAC learnability. The accuracy parameter ϵ determines how far we allow our predictor h to be from the optimal predictor ("approximately correct"), and the confidence parameter δ indicates how likely h is to meet the accuracy requirement ("probably").

Remark 4.1.5. Definition 4.1.4 does not imply anything on the computational aspects of learnability. Nonetheless:
(i) Some definitions require $m(\epsilon, \delta)$ to grow polynomially with its parameters as they tend to 0, which, for example, does not allow $m(\epsilon, \delta) = \epsilon^{-1} 2^{1/\delta}$.
(ii) Some definitions include the number of computations of $\mathcal{A}$ as a parameter of m (and ask for polynomial growth).

Remark 4.1.6. The PAC learning definition is not easy to digest at first. After some time, we may even wonder if the definition is not vacuous. Indeed, since m can grow at any pace in terms of its parameters and there is no restriction either on $\mathcal{A}_{\text{ERM}}$, we can set $m(\epsilon, \delta)$ extremely large (think of $10^{(\epsilon\delta)^{-1000000}}$). By the law of large numbers and the central limit theorem (Theorems 2.8.1 and 2.8.2) we know that $L(h) \to R_D(h)$ as $m \to \infty$ with an approximation error of order $O(1/\sqrt{m})$, and since the realisability assumption holds, it feels like we should always be likely to find a good enough predictor. However, this intuition does not take into account the following fact: $m(\epsilon, \delta)$ and $\mathcal{H}$ are fixed **before** choosing f and D. In Corollary 6.4 and Theorem 5.1 in [39], the interested reader will find a construction of a non-PAC learnable hypothesis class. We now **informally** explain the main idea on a particular case.

Let $\mathcal{H}$ be the set of all functions from $\mathbb{R}$ to $\{0, 1\}$; we claim that it is not PAC learnable, and to demonstrate this, as soon as we choose $m(\epsilon, \delta)$, we adversarially construct D and f as follows. After we fix $m(\epsilon, \delta)$, we choose arbitrary $2m(\epsilon, \delta)$ pairwise distinct points in $\mathbb{R}$. We let D be the uniform discrete distribution on these $2m(\epsilon, \delta)$ points. Because $\mathcal{H}$ contains all $\{0, 1\}$-binary functions, we can label them in any possible way with an $f \in \mathcal{H}$ so that the realisability assumption holds. Because (at least) half of the $2m(\epsilon, \delta)$ points will not be in the dataset S (recall that it is sampled with D), an ERM predictor $h_{\mathcal{A}}$ with $L(h_{\mathcal{A}}) = 0$ cannot have learned anything on half of the points and therefore will be likely not ϵ-close to f. Conclusion: $\mathcal{H}$ is not PAC learnable.

4.1.3 Agnostic PAC learnability

We saw that PAC learning offers guarantees on a hypothesis class that enables the retrieval of a labeling function **within that class**, since Definition 4.1.4 assumes the realisability assumption. However, the datasets a practitioner meets are not, in general, labeled by a function that belongs to the hypothesis class they use. Even in the case of a truly linear relationship between x and y, it is enough, for example, to have noise in the samples so that the realisability assumption does not hold for the set of linear predictors. This means that we cannot guarantee that the generalization error $R_D(h) = 0$; we still want to find $h \in \mathcal{H}$ for which the $R_D(h)$ is nevertheless low.

It turns out that the realisability assumption can be removed and the PAC learning formalism can be extended. Recall that $\mathcal{A}_{\text{ERM}}$ denotes an algorithm that returns a predictor $h_{\mathcal{A}} \in \arg\min_{h \in \mathcal{H}} L(h)$. Note, however, that since we removed the realisability assumption, we no longer have the guarantee that $L(h_{\mathcal{A}}) = 0$.

Definition 4.1.7 (Agnostic PAC learnability). A hypothesis class $\mathcal{H}$ is agnostic PAC learnable if there exists a function $m_{\mathcal{H}} : (0, 1)^2 \to \mathbb{N}$ with the following property: for all $\epsilon, \delta \in (0, 1)$, every distribution $\mathcal{D}$ over $\mathcal{X}$, and every labeling function $f : \mathcal{X} \to \{0, 1\}$, when running $\mathcal{A}_{\text{ERM}}$ on $m \geq m_{\mathcal{H}}(\epsilon, \delta)$ i. i. d. samples, the hypothesis $h_{\mathcal{A}}$ is such that with probability at least $1 - \delta$,

$$R_D(h_A) \le \min_{h^* \in \mathcal{H}} R_D(h^*) + \epsilon.$$

Remark 4.1.8. When the realisability assumption does not hold, no learner can guarantee an arbitrarily small error. Under the definition of agnostic PAC learning, a learner can still declare success if its error is not much larger than the best error achievable by a predictor from the class $\mathcal{H}$. This is in contrast to PAC learning, in which the learner is required to achieve a small error in absolute terms and not relative to the best error achievable by the hypothesis class. In particular, for a given task, a hypothesis class $\mathcal{H}$ could be a very poor choice and still be agnostic PAC learnable. More informally and concisely, agnostic PAC learnable does **not** imply a good model choice.

4.2 Finite-sized hypothesis classes

The PAC learning formalism tells us that if $\mathcal{H}$ is too complex, then we may not be able to find good predictors in $\mathcal{H}$ from finitely many data samples. The first restriction we consider is when $\mathcal{H}$ is finite, that is, $\mathcal{H}$ contains only a finite number of functions. Is $\mathcal{H}$ simple enough to be PAC learnable? Throughout this section, we work under the assumption that $\mathcal{H}$ is finite.

4.2.1 PAC learnability

It turns out that any finite hypothesis class is PAC learnable:

Theorem 4.2.1. *Suppose that the realisability assumption holds. If $\mathcal{H}$ is finite, then it is PAC learnable. Moreover, the map $m : (0,1)^2 \to \mathbb{N}$ in Definition 4.1.4 can be chosen as*

$$m(\epsilon, \delta) = \left\lceil \frac{\log(|\mathcal{H}|/\delta)}{\epsilon} \right\rceil,$$

where $\lceil x \rceil$ denotes the smallest integer greater than or equal to $x \in \mathbb{R}$.

Proof. Let h_A denote the predictor trained with algorithm A_{ERM}. Note that h_A depends on the dataset S; if we sample m i. i. d. data points according to D, then we look at S as a random variable in $(\mathcal{X} \times \mathcal{Y})^m$ with law D^m.

Fix $\epsilon, \delta \in (0,1)$ and let the map $m(\cdot, \cdot)$ be defined as in the theorem. Proving that $\mathcal{H}$ is PAC learnable amounts to proving that

$$D^{m(\epsilon, \delta)}(S : R_D(h_A) > \epsilon) < \delta. \tag{4.1}$$

We will make use of the three following basic facts that we admit without proof:
(i) For all $x \in \mathbb{R}, 1 + x \le e^x$.
(ii) For all sets $A, B \in \mathcal{F}$ and probability P, $P(A \cup B) \le P(A) + P(B)$.
(iii) If $A \subset B$, then $P(A) \le P(B)$.

Fix $m \in \mathbb{N}$ and let us bound $D^m(S : R_D(h_A) > \epsilon)$. Define the set of bad predictors

$$\mathcal{H}_b := \{h \in \mathcal{H} : R_D(h) > \epsilon\}$$

and define the set of misleading datasets of size m

$$M_m := \{S \in (\mathcal{X} \times \mathcal{Y})^m : \exists h \in \mathcal{H}_b, L(h) = 0\}.$$

We use "misleading" to stress that even though the hypothesis class contains a predictor with null empirical error, this predictor fails to achieve a generalization smaller than our threshold ϵ. Finally, for all $h \in \mathcal{H}_b$, define the set of misleading datasets of size m for h by

$$M_m(h) := \{S \in (\mathcal{X} \times \mathcal{Y})^m : L(h) = 0\}.$$

Note that by definition we can write

$$M_m = \bigcup_{h \in \mathcal{H}_b} M_m(h).$$

In particular, by fact (ii) we have that

$$D^m(M_m) \le \sum_{h \in \mathcal{H}_b} D^m(M_m(h)).$$

Because the m samples of the dataset are independent, for all $h \in \mathcal{H}_b$, we have that

$$D^m(M_m(h)) = D^m(S \in: h(x_i) = y_i, \quad \forall i \in \{1, \ldots, m\}) = \prod_{i=1}^m D(x_i : h(x_i) = y_i),$$

and by the definition of the generalization error,

$$R_D(h) = \mathbb{E}_{x \sim D}(\mathbb{1}_{\{h(x) \neq y\}}) = D(x_i : h(x_i) \neq y_i).$$

Since $h \in \mathcal{H}_b$, we moreover have that

$$D(x_i : h(x_i) = y_i) = 1 - R_D(h) \le 1 - \epsilon.$$

In particular, by fact (i) we see that

$$D^m(M_m(h)) \le (1 - \epsilon)^m \le e^{-\epsilon m}.$$

Putting everything together, we have shown that

$$D^m(M_m) \le \sum_{h \in \mathcal{H}_b} e^{-\epsilon m} \le |\mathcal{H}| e^{-\epsilon m}.$$

Recall that our goal is to establish (4.1), which is not equivalent to the above bound. Indeed, we just upper bounded **the sum over bad hypotheses** of the probability to sample a misleading dataset. Nonetheless, the probability of sampling a misleading dataset for the trained predictor (as in (4.1)) is upper bounded by the left hand-side above, thanks to fact (iii): since $\{S \in (\mathcal{X} \times \mathcal{Y})^m : R_D(h_A) > \epsilon\} \subset \bigcup_{h \in \mathcal{H}_b} \{S \in (\mathcal{X} \times \mathcal{Y})^m : R_D(h) > \epsilon\} = M_m$, we have that

$$D^m(S : R_D(h_A) > \epsilon) \leq D^m(M) \leq |\mathcal{H}|e^{-\epsilon m}.$$

It suffices now to check that for $m > m(\epsilon, \delta) = \frac{\log(|\mathcal{H}|/\delta)}{\epsilon}$, the right hand-side above is smaller than δ, which is the case, hence concluding the proof. □

Given a finite hypothesis class and two numbers $\epsilon, \delta \in (0, 1)$, Theorem 4.2.1 provides a sufficient number of data points to learn a good predictor for binary classification, that is to say, with generalization error smaller than ϵ and with probability greater than $1 - \delta$, even in the worst-case scenario where the labeling function (in $\mathcal{H}$) and the data distribution are chosen adversarially.

We can analyze the bound of Theorem 4.2.1 and note that:

- As $m \to \infty$, the probability of picking a misleading set for which $h \in \mathcal{H}_b$ yields $L(h) = 0$ decreases.
- As $|\mathcal{H}|$ increases, so does the (adversarial) probability of finding bad hypothesis.
- The smaller we choose ϵ (i.e., we want better accuracy), the greater we need to choose m.

Remark 4.2.2. It is important to note that Theorem 4.2.1 does not say that if $m < m(\epsilon, \delta)$, then training will fail. In other words, the bound is not tight. In practical problems, it is uncommon to know whether $m > m(\epsilon, \delta)$ or not.

4.2.2 Agnostic PAC learnability

Recall Definition 4.1.7, where agnostic PAC learnability was defined. We proved in Theorem 4.2.1 that any finite hypothesis class is PAC learnable and derived a bound for the number of data samples that guarantee success of training. We are now interested in removing the realisability assumption.

Theorem 4.2.3. *If $\mathcal{H}$ is finite, then it is agnostic PAC learnable. Moreover, the map $m : (0, 1)^2 \to \mathbb{N}$ in Definition 4.1.7 can be chosen as*

$$m(\epsilon, \delta) := \left\lceil \frac{2\log(2|\mathcal{H}|/\delta)}{\epsilon^2} \right\rceil.$$

Before we prove Theorem 4.2.3, we need to introduce the uniform convergence property.

Definition 4.2.4. We say that a hypothesis class $\mathcal{H}$ has the uniform convergence property if there is a function $m^{UC} : (0,1)^2 \rightarrow \mathbb{N}$ such that for all $\epsilon, \delta \in (0,1)$ and every probability distribution D over $\mathcal{X}$, if $m \geq m^{UC}(\epsilon, \delta)$ and the m data samples in S are sampled i. i. d. with common law D, then

$$D^m(\exists h \in \mathcal{H} : |R_D(h) - L(h)| > \epsilon) < \delta.$$

In short, the uniform convergence property states that provided that m is large enough, the generalization gap is likely to be small across all data distribution D and labeling function f.

Proof of Theorem 4.2.3. The strategy is the following:
- We show that every finite $\mathcal{H}$ has the uniform convergence property of Definition 4.2.4.
- We show that if $\mathcal{H}$ has the uniform convergence property, then it is agnostic PAC learnable with $m(\epsilon, \delta) = m^{UC}(\epsilon/2, \delta)$.

Let $\mathcal{H}$ be a finite hypothesis class. To show that it has the uniform convergence property, we admit without proof that

$$D^m(|R_D(h) - L(h)| > \epsilon) \leq 2e^{-2m\epsilon^2} \quad \forall h \in \mathcal{H}.$$

It is a simple application of Hoeffding's inequality (see, e. g., [6]) that provides a bound on the probability that an empirical mean is far from its expectation. (Note that the expectation of $L(h)$ integrated over the datasets under D^m is indeed given by $R_D(h)$.) We now use fact (ii) from the proof of Theorem 4.2.1 to write

$$D^m\left(\bigcup_{h \in \mathcal{H}} \{|R_D(h) - L(h)| > \epsilon\} \right) \leq \sum_{h \in \mathcal{H}} 2e^{-2m\epsilon^2} \leq |\mathcal{H}|2e^{-2m\epsilon^2}.$$

Hence we can choose $m^{UC}(\epsilon, \delta) = \frac{\log(2|\mathcal{H}|/\delta)}{2\epsilon^2}$, and we see that $\mathcal{H}$ has the uniform convergence property.

We now show that the uniform convergence property implies the agnostic PAC learning property. Let h_A denote the predictor obtained by the training algorithm, and let h^* be the optimal predictor within the class, that is,

$$h^* := \arg\min_{h \in \mathcal{H}} R_D(h).$$

Define the event $E_{unif} := \{\forall h \in \mathcal{H} : |R_D(h) - L(h)| < \epsilon\}$. On the event E_{unif}, we have that

$$R_D(h_A) \leq L(h_A) + \epsilon \leq L(h^*) + \epsilon \leq R_D(h^*) + 2\epsilon,$$

where we used that by definition h_A achieves minimal empirical loss. Note that the uniform convergence property ensures that $D^m(E_{\text{unif}}) > 1-\delta$, which means in particular that the above inequalities hold with probability greater than $1 - \delta$.

This shows that setting $m(\epsilon, \delta) := m^{\text{UC}}(\epsilon/2, \delta)$, $\mathcal{H}$ satisfies the agnostic PAC learnability property. More explicitly,

$$m(\epsilon, \delta) = \frac{2\log(2|\mathcal{H}|/\delta)}{\epsilon^2},$$

as claimed, which concludes the proof. □

Remark 4.2.5 (Infinite-sized hypothesis classes). We have assumed that $|\mathcal{H}|$ is finite and showed that finite classes are learnable and that the sample complexity of a hypothesis class is upper bounded by an expression that involves the log of its size. What happens when $|\mathcal{H}| = \infty$? We want to say something about the expressiveness of a set of functions.

Let us consider a hypothesis class $\mathcal{H}$ that contains all functions

$$f : S \to \{-1, 1\}^{|S|}.$$

This means that this hypothesis class is capable of correctly labeling all datasets of size $|S|$.

The Vapnik–Chervonenkis dimension (VC-dimension) of a hypothesis class $\mathcal{H}$ measures precisely the following: it is the maximal size of a set S that can always be correctly labeled by elements in $\mathcal{H}$. For example, $\dim_{\text{VC}}(\{h = \mathbb{1}_{[a,\infty)}; a \in \mathbb{R}\}) = 1$ and $\dim_{\text{VC}}(\{h = \mathbb{1}_{[a,b]}; a \leq b \in \mathbb{R}\}) = 2$ (note that both classes contain infinitely many hypotheses). For more information on this, see [46].

4.3 Bias-complexity tradeoff and bias-variance tradeoff

4.3.1 Existence of noise

Consider again the setting of classification with $\mathcal{Y} = \{-1, 1\}$. Let us introduce the notion of noisy labels.

Suppose we have a dataset $S = \{(x_1, y_1), \ldots, (x_n, y_n)\}$. We now suppose that each data point and its corresponding label are independently drawn from an unknown data distribution, i.e., $(x_i, y_i) \sim \mathcal{D}(X, Y)$. Note the difference with the setup where the labels are given by a deterministic labeling function f: here the same x can have a positive probability to be labeled by -1 or 1.

This lack of a perfect labeling function f accounts for the noise on the label. For example, we can think of this as, for example, the features do not contain all the information needed to attribute the label in a deterministic way. This setting is a bit closer to reality, because of maybe lack of information, noise, or other source of uncertainty, the labeling function may be nondeterministic.

Example 4.3.1. Suppose we have a model for tastiness of a papaya given the color and softness. Suppose that most soft, brightly colored papayas are tasty. However, we can have the situation that the papaya is soft and bright but still not tasty (e. g., because of bad climate), even if it is unlikely.

Under this new assumption that there is also noise on the labels, we can write a theoretical optimal classifier.

Definition 4.3.2 (Bayes optimal predictor). Given a probability distribution $\mathcal{D}$ over $\mathcal{X} \times \mathcal{Y}$, the predictor is defined as

$$f_{\text{Bayes}} = \begin{cases} 1 & \text{if } \mathcal{D}(y = 1|x) \geq 1/2, \\ -1 & \text{otherwise.} \end{cases}$$

- In the deterministic case, $\mathcal{D}(y = 1|x)$ is either 1 or 0, because we have a deterministic map f.
- Under uncertainty, on average, this predictor is optimal:

$$R_D(f_{\text{Bayes}}) \leq R_D(h) \quad \forall h : \mathcal{X} \to \mathcal{Y}.$$

- We rarely have access to this classifier, because it implies that we can evaluate the probability $\mathcal{D}(y = 1|x)$. Typically,

$$f_{\text{Bayes}} \notin \mathcal{H}.$$

The error made by f_{Bayes} by definition is

$$R_D(f_{\text{bayes}}) = \mathbb{E}_x[1 - \max\{(\mathcal{D}(-1|x), \mathcal{D}(1|x))\}],$$

and this is the minimal theoretical error possible. This leads us to defining the *noise* as follows.

Definition 4.3.3. Given a distribution $\mathcal{D}$ over $\mathcal{X} \times \mathcal{Y}$, the noise at a point $x \in \mathcal{X}$ is defined as

$$\text{noise}(x) = 1 - \max\{(\mathcal{D}(-1|x), \mathcal{D}(1|x))\}.$$

The noise is a characteristic of the learning task and indicates its level of difficulty. For example, for a point x, if the noise is 1/2, then it will be challenging to predict its label correctly. On the contrary, a noise of 0 means that there exists a labeling function.

4.3.2 Bias-complexity trade-off

In learning theory, we talk about bias-complexity trade-off. The generalization error can be decomposed into two components (three components if we include noise):

$$R_D(h) = \epsilon_{\text{approx}} + \epsilon_{\text{est}}, \tag{4.2}$$

where $\epsilon_{\text{approx}} = \min_{h' \in \mathcal{H}} R_D(h')$ and $\epsilon_{\text{est}} = R_D(h) - \epsilon_{\text{approx}}$.

Approximation error:
this is the minimum generalization error achievable by a predictor in the hypothesis class $\mathcal{H}$. This term measures how much error we have because we restrict ourselves to a specific class, that is, how much *inductive bias* we have. This error does not depend on the sample size and is determined by the hypothesis class chosen. Enlarging the hypothesis class (e. g., making it more complicated) can only decrease the approximation error. Note that under the **realisability assumption**, the approximation error is zero, whereas in the agnostic case, it can be large.

Example 4.3.4. Using a finite polynomial basis to represent a nonpolynomial function induces an inherent approximation error.

Estimation error:
Consider an ERM predictor $h_* \in \arg\min_{h \in \mathcal{H}} L(h)$. The estimation error is the gap between h_* and the best predictor in $\mathcal{H}$ in terms of generalization error. This is in general nonnull because the empirical error is only an estimate of the generalization error, and therefore h_* does not necessarily reach the minimal generalization error over the hypothesis set. This quantity depends on the **training set size** and on **the size and complexity of the hypothesis class** (namely, ϵ_{est} increases logarithmically with size of $\mathcal{H}$ and decreases with m increasing).

Since we want to minimize the total error, we have a trade-off, called a **bias-complexity trade-off**. A rich $\mathcal{H}$ reduces the approximation error but may lead to high estimation error (overfitting), whereas a simple $\mathcal{H}$ reduces the estimation error but increases the approximation error (underfitting).

4.3.3 Bias–variance trade-off

The *Bias–Variance trade-off* is typically referenced in computational statistics and does not use the learning framework we have described in this Chapter (PAC). Nevertheless, it is used quite often to describe the notion of managing the complexity of a chosen model class with its different sources of errors. Bias and variance are schematically illustrated in Figure 4.1.

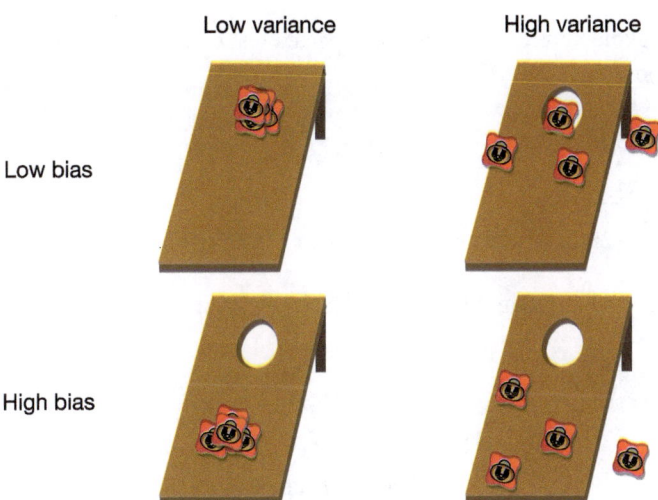

Figure 4.1: Graphical representation of bias and variance (precision and accuracy).

In this section, we leave the binary classification setup to consider a regression task with the mean squared error. In the theorem below, we consider that in the dataset, there is noise of the following form: for some unknown map $f : \mathcal{X} \to \mathcal{Y}$, the dataset $S = \{(x_i, y_i); i = 1, \ldots, m\}$ is such that for all $i = 1, \ldots, m$, $y_i = f(x_i) + \epsilon_i$, where the law of ϵ_i depends on x_i and has bounded variance.

We denote by $\mathcal{D}$ the law of a pair (x_i, y_i) and by $\mathcal{D}^m$ the law of S composed of m i. i. d. samples under $\mathcal{D}$. To make the sources of randomness explicit, we write $\mathbb{E}_z[\cdot]$ for the expectation where all variables z are integrated; for example, $\mathbb{E}_{(x,y)}[(g(x) - y)^2]$ means that x and y are random variables and the expectation is over $(x,y) \sim \mathcal{D}$. Similarly, $\mathbb{E}_{z|z'}[\cdot]$ denotes the conditional expectation where z is integrated conditionally given z'.

Theorem 4.3.5. *Consider the setting described above. For a predictor $h_S \in \arg\min_{h \in \mathcal{H}} L(h)$, the expected mean-squared error can be decomposed as*

$$\underbrace{\mathbb{E}_{x,y,S}[(h_S(x) - y)^2]}_{\text{Expected Test Error}} = \underbrace{\mathbb{E}_{x,S}[(h_S(x) - \bar{h}(x))^2]}_{\text{Variance}} + \underbrace{\mathbb{E}_{x,y}[(\bar{y}(x) - y)^2]}_{\text{Noise}} + \underbrace{\mathbb{E}_x[(\bar{h}(x) - \bar{y}(x))^2]}_{\text{Bias}^2},$$

$$\bar{h}(\cdot) = \mathbb{E}_S[h_S(\cdot)],$$

and

$$\bar{y}(x) = \mathbb{E}_{y|x}[y] = \mathbb{E}_{\epsilon|x}[f(x) + \epsilon].$$

Note that if the noise is centered, then we get $\bar{y}(x) = f(x)$, and the noise term in Theorem 4.3.5 vanishes. Furthermore, letting $g(x) := \mathbb{E}_{\epsilon|x}[\epsilon]$ be the expectation of the noise given x, we can always define $\tilde{f} := f + g$ to be the function to approximate, which centers the noise $\tilde{\epsilon} = \epsilon - g(x)$.

Proof. Artificially adding and subtracting in $\bar{h}$, the expected test error of the ERM predictor h_S can be decomposed as

$$\mathbb{E}_{(x,y),S}[(h_S(x) - y)^2] = \mathbb{E}_{(x,y),S}[(h_S(x) - \bar{h}(x) + \bar{h}(x) - y)^2]$$
$$= \mathbb{E}_{x,S}[(h_S(x) - \bar{h}(x))^2]$$
$$+ 2\mathbb{E}_{(x,y),S}[(h_S(x) - \bar{h}(x))(\bar{h}(x) - y)] + \mathbb{E}_{x,y}[(\bar{h}(x) - y)^2]. \quad (4.3)$$

The middle term of the above equation is 0, since basic properties of conditional expectation yield that

$$\mathbb{E}_{(x,y),S}[(h_S(x) - \bar{h}(x))(\bar{h}(x) - y)] = \mathbb{E}_{(x,y)}[\mathbb{E}_{S|(x,y)}[h_S(x) - \bar{h}(x)](\bar{h}(x) - y)]$$
$$= \mathbb{E}_{(x,y)}[(\mathbb{E}_{S|(x,y)}[h_S(x)] - \bar{h}(x))(\bar{h}(x) - y)]$$
$$= \mathbb{E}_{(x,y)}[(\bar{h}(x) - \bar{h}(x))(\bar{h}(x) - y)]$$
$$= \mathbb{E}_{x,y}[0]$$
$$= 0.$$

Returning to (4.3), we are left with two terms:

$$\mathbb{E}_{(x,y),S}[(h_S(x) - y)^2] = \underbrace{\mathbb{E}_{x,S}[(h_S(x) - \bar{h}(x))^2]}_{\text{Variance}} + \mathbb{E}_{(x,y)}[(\bar{h}(x) - y)^2]. \quad (4.4)$$

Artificially adding and subtracting $\bar{y}(x)$ in the second term of the right-hand side, we get

$$\mathbb{E}_{(x,y)}[(\bar{h}(x) - y)^2] = \underbrace{\mathbb{E}_{(x,y)}[(\bar{y}(x) - y)^2]}_{\text{Noise}} + \underbrace{\mathbb{E}_x[(\bar{h}(x) - \bar{y}(x))^2]}_{\text{Bias}^2}$$
$$+ 2\,\mathbb{E}_{(x,y)}[(\bar{h}(x) - \bar{y}(x))(\bar{y}(x) - y)].$$

The third term in the right-hand side above is 0, since

$$\mathbb{E}_{(x,y)}[(\bar{h}(x) - \bar{y}(x))(\bar{y}(x) - y)] = \mathbb{E}_x[\mathbb{E}_{y|x}[\bar{y}(x) - y](\bar{h}(x) - \bar{y}(x))]$$
$$= \mathbb{E}_x[(\bar{y}(x) - \bar{y}(x))(\bar{h}(x) - \bar{y}(x))]$$
$$= \mathbb{E}_x[0]$$
$$= 0.$$

This gives us the decomposition of expected test error

$$\underbrace{\mathbb{E}_{x,y,S}[(h_S(x) - y)^2]}_{\text{Expected Test Error}} = \underbrace{\mathbb{E}_{x,S}[(h_S(x) - \bar{h}(x))^2]}_{\text{Variance}} + \underbrace{\mathbb{E}_{x,y}[(\bar{y}(x) - y)^2]}_{\text{Noise}} + \underbrace{\mathbb{E}_x[(\bar{h}(x) - \bar{y}(x))^2]}_{\text{Bias}^2},$$

as claimed, which concludes the proof. □

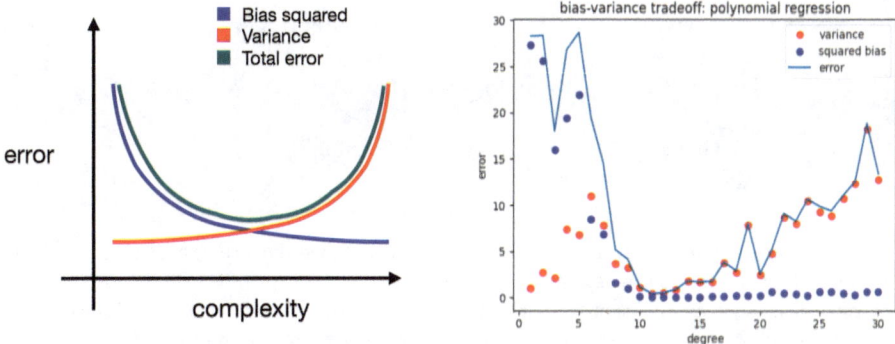

Figure 4.2: Left: Schematic of the contributions of variance and squared bias to the total error as a function of the complexity of the model. Right: Numerical experiment where complexity is the degree used in polynomial regression to fit a noisy function. We observe a "U shape" for the error, whereas the contribution of the variance increases, and that of the squared bias decreases.

In words, Theorem 4.3.5 decomposes the mean-squared test error of the ERM predictor into the sum of

the variance, which captures how "overspecialized" your model is to a particular training set (i. e., overfitting),

the bias, which is the inherent error of our model even with infinite training data (this is because our model is "inherently biased" to a particular kind of solution, e. g., a linear function), and

the noise, which measures the shift in the data due to intrinsic uncentered noise (we can never beat this; it is an aspect of the data).

The tradeoff between the bias and the variance is illustrated in Figure 4.2.

Remark 4.3.6. What about neural networks? Do not they have millions (sometimes billions) of parameters and are still able to generalize? The bias–variance trade-off implies that a model should balance underfitting and overfitting: rich enough to express underlying structure in data and simple enough to avoid fitting spurious patterns. However, in modern practice, very rich models such as neural networks are trained to exactly fit (i. e., interpolate) the data. Classically, such models would be considered overfitting, and yet they often obtain high accuracy on test data. In [4] the authors show a "double-descent" curve (Figure 4.3), which includes the textbook U-shaped bias–variance trade-off curve by showing how increasing model capacity (in some model types) beyond the point of interpolation results in improved performance. This is observed, for instance, in neural networks and ensemble methods with decision trees. This has also been observed in numerical experiments (see Figure 4.4).

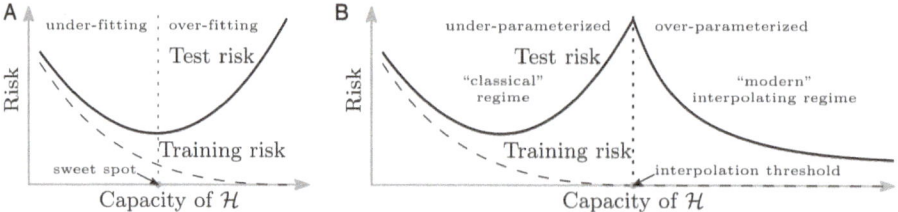

Figure 4.3: Curves for training risk and test risk. (A) The classical U-shaped risk curve arising from the bias–variance trade-off. (B) The double-descent risk curve, which incorporates the U-shaped risk curve (i. e., the *classical* regime) together with the observed behavior from using high-expressivity function classes (i. e., the *modern* interpolating regime), separated by the interpolation threshold. The predictors to the right of the interpolation threshold have zero training risk. Source: [5].

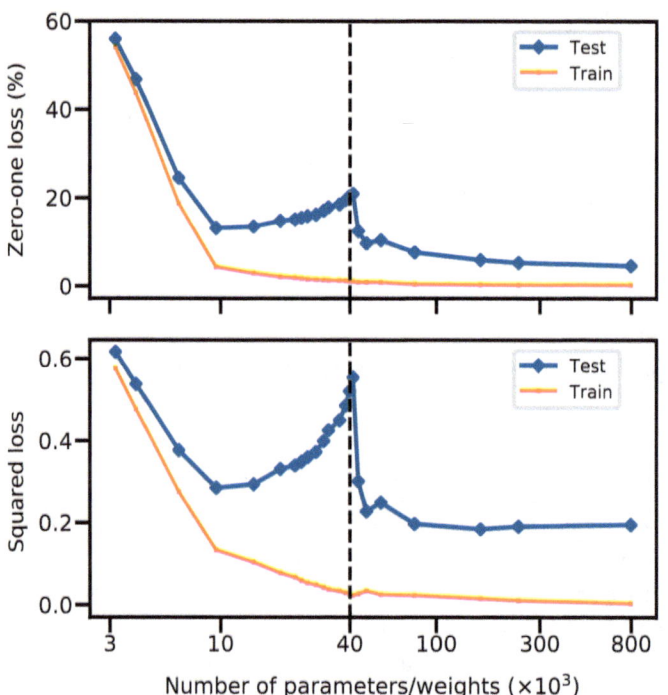

Figure 4.4: Double-descent risk curve for a fully connected neural network on MNIST. Training and test errors are shown for different losses. The dataset considered has 4000 datapoints, with feature dimension $d = 784$ and $K = 10$ classes. The number of parameters of the network is given by $(d + 1)H + (H + 1)K$. The interpolation threshold (black dashed line) is observed at $n \cdot K$. Source: [5].

5 Linear models

In this chapter, we study the family of *linear predictors*. Linear predictors are intuitive, easy to interpret, theoretically sound, and fit the data reasonably well in many natural learning problems.

Let $\mathcal{X} = \mathbb{R}^d$ and $\mathcal{Y} = \mathbb{R}$. We consider the parametric class of affine functions

$$\mathcal{H} = \{f_w : x \to f_w(x) = \vec{w}^T \vec{x} + b : \vec{w} \in \mathbb{R}^d, b \in \mathbb{R}\}, \tag{5.1}$$

where each function is parameterized by $\vec{w}$ and b, takes as input a vector $\vec{x} \in \mathbb{R}^d$, and returns a scalar $\vec{w}^T \vec{x} + b$. We could consider the output space to be $\mathbb{R}^{d'}$ for $d' \geq 2$ without changing the theory but making the presentation slightly more difficult. We will stick to $\mathcal{Y} = \mathbb{R}$ throughout the chapter.

We call the **homogeneous representation** of the parameters the concatenated form $\vec{w} = (w_1, \ldots, w_d, b)$. The class above is then equivalent to setting the input space to be $\mathbb{R}^d \times \{1\}$ and $f_w(\vec{x}) = \vec{w}^T \vec{x}$.

Even though the output space is $\mathbb{R}$, linear models are not limited to regression tasks: for examples for binary classification, we can compose an element of $\mathcal{H}$ with the sign function, which returns the sign of a real number (positive or negative); see Section 5.2.

5.1 Linear regression

5.1.1 Cost function choice

Linear predictors are nice because they are linear both in the inputs and in the parameters and thus preserve nice properties of the loss function $\ell : \mathcal{Y} \times \mathcal{Y} \to \mathbb{R}_+$ to the (empirical) *cost function* $C : \mathbb{R}^d \to \mathbb{R}_+$, defined on the parameter space by

$$C(w) := \frac{1}{m} \sum_{i=1}^{m} \ell(f_w(x_i), y_i). \tag{5.2}$$

For example, since the composition of differentiable and convex maps is differentiable and convex, if ℓ is convex in its first argument, then so is C on the parameter space. In particular, **training on $\mathcal{H}$ with gradient flow on the parameters is guaranteed to converge to a global minimum of the cost function** (see Chapter 3).

Common choices of loss functions are
- the squared error loss (or L_2 loss) $\ell : (y, y') := \frac{1}{2}(y - y')^2$,
- the absolute error loss (or L_1 loss) $\ell : (y, y') \mapsto |y - y'|$, and
- the Huber loss

$$\ell(y, y', \delta) = \begin{cases} \frac{1}{2}(y - y')^2 & \text{for } |y - y'| < \delta, \\ \delta(|y - y'| - \frac{1}{2}\delta) & \text{otherwise.} \end{cases}$$

https://doi.org/10.1515/9783111288994-005

The squared error loss is nice for theoretical reasons: it is convex and differentiable and has no hyperparameters, but it is not robust to outliers (i. e., the total error may be dominated by a point that is an outlier, since the difference grows squared). On the contrary, the absolute error loss is less prone to potential outliers (see Figure 5.1, right) and is convex too. However, it is not differentiable at 0. The Huber loss is a compromise between squared and absolute error losses: it is convex and both differentiable and robust to outliers. The price to pay is that it has a hyperparameter δ that has to be tuned.

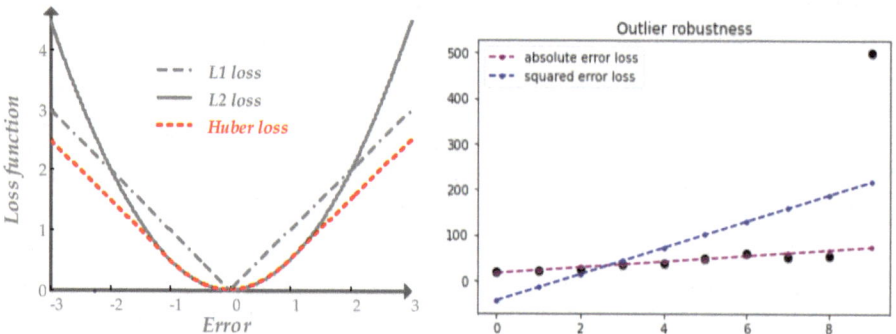

Figure 5.1: *Left*: Different shapes of loss functions. *Right*: Example of an optimal linear predictor for the squared and absolute error losses with the presence of outlier.

5.1.2 Explicit solution

Least squares is the method that solves the empirical risk minimization problem for the hypothesis class (5.1) with respect to the squared loss. We want to find w that minimizes

$$\arg\min_{w} C(w) = \arg\min_{w} L(f_w) = \arg\min_{w} \frac{1}{2m} \sum_{i=1}^{m} (w^T x_i - y_i)^2.$$

Note that here we use the homogeneous notation: $w = (w_1, \ldots, w_n, b)$, $x_i = (x_{i1}, \ldots, x_{in}, 1)^T$. We will use the more compact notation and equivalent formulation

$$\arg\min_{w} C(w) = \frac{1}{2m} \arg\min_{w} \|Xw - Y\|^2, \tag{5.3}$$

where $X = (x_{ij})_{ij} \in \mathbb{R}^{m \times n}$, $Y = (y_1, \ldots, y_m)^T$, and $\| \cdot \|$ is the Euclidean norm in $\mathbb{R}^m$. The number m is the number of samples, and n is the number of *features*.

Definition 5.1.1 (Moore–Penrose pseudoinverse). Let $A \in \mathbb{R}^{m \times n}$, a matrix $A^+ \in \mathbb{R}^{n \times m}$ is a Moore-Penrose pseudoinverse of a real matrix A if it satisfies the following conditions:

1. $AA^+A = A$.
2. $A^+AA^+ = A^+$.
3. $(AA^+)^T = AA^+$.
4. $(A^+A)^T = A^+A$.

It can be shown that for any matrix $A \in \mathbb{R}^{m \times n}$, there exists a unique Moore–Penrose inverse: use the singular value decomposition $A = U\Sigma V^T$ and check that $A^+ = V\Sigma^{-1}U^T$ satisfies the above definition, where Σ is an $m \times n$ rectangular diagonal matrix, and Σ^{-1} is an $n \times m$ rectangular diagonal matrix with diagonal entries the inverses of those of Σ when positive and 0 when the corresponding entry in Σ is 0.

The Moore–Penrose pseudoinverse of a matrix generalizes the inverse to noninvertible matrices. The geometric interpretation is the following: for any matrix $X \in \mathbb{R}^{m \times n}$, denote $\mathrm{Ker}(X) := \{u \in \mathbb{R}^n : Au = 0\}$ and $\mathrm{Im}(X) = \{v \in \mathbb{R}^m : \exists u \in \mathrm{Ker}(X)^{\perp}, Xu = v\}$. Then for all elements $v \in \mathrm{Im}(X)$, there exists a unique element $u \in \mathrm{Ker}(X)^{\perp}$ such that $Xu = v$ (otherwise, if there were two such u and u', then we would have $X(u - u') = 0$ and thus $u - u' \in \mathrm{Ker}(X) \cap \mathrm{Ker}(X)^{\perp} = \{0\}$). The pseudoinverse $X^+ \in \mathbb{R}^{n \times m}$ returns the following: for any $v \in \mathbb{R}^m$, consider its unique orthogonal projection $\pi(v)$ onto $\mathrm{Im}(X)$; then X^+v is the unique $u \in \mathrm{Ker}(X)^{\perp}$ such that $Xu = \pi(v)$. In other words, the restriction of X^+ on $\mathrm{Im}(X)$ is the inverse of the restriction of X on $\mathrm{Ker}(X)^{\perp}$.

Indeed, for every $v \in \mathrm{Im}(X)$, there exists a unique $u \in \mathrm{Ker}(X)^{\perp}$ such that $Xu = v$. Then by item 2 of Definition 5.1.1 we see that $XX^+v = XX^+Xu = Xu = v$. We can reason similarly with item 1 and show that for every $u \in \mathrm{Ker}(X)^{\perp}$, there exists a unique $v \in \mathrm{Im}(X)$ such that $X^+v = u$.

We deduce the following lemma.

Lemma 5.1.2. *For all $v \in \mathbb{R}^m$, we have*

$$v = \underbrace{XX^+v}_{\in \mathrm{Im}(X)} + \underbrace{(\mathcal{I}_m - XX^+)v}_{\in \mathrm{Im}(X)^{\perp}}.$$

Similarly, for all $u \in \mathbb{R}^n$, we have

$$u = \underbrace{X^+Xu}_{\in \mathrm{Ker}(X)^{\perp}} + \underbrace{(\mathcal{I}_m - X^+X)u}_{\in \mathrm{Ker}(X)}.$$

Proof. It is clear that the two sides of the equation are equal. It is also clear that $XX^+v \in \mathrm{Im}(X)$, since it is X applied to the vector X^+v. It is a standard linear algebra fact that there exists a unique decomposition of v into the sum of an element of $\mathrm{Im}(X)$ and an element of $\mathrm{Im}(X)^{\perp}$. Therefore we only need to show that $(\mathcal{I}_m - XX^+)v \in \mathrm{Im}(X)^{\perp}$. Since the restriction of X on $\mathrm{Ker}(X)^{\perp}$ has inverse the restriction of X^+ on $\mathrm{Im}(X)$, we only need to check that by applying X^+ to $(\mathcal{I}_m - XX^+)v$ we obtain 0:

$$X^+(\mathcal{I}_m - XX^+)v = (X^+ - X^+XX^+)v = 0$$

by item 2 of Definition 5.1.1. This shows the first claim. The second claim follows from an identical argument, and this concludes the proof. $\square$

Theorem 5.1.3. *Let $X \in \mathbb{R}^{m \times n}$ and $Y \in \mathbb{R}^m$, and let X^+ denote the Moore–Penrose pseudoinverse. Then w is a solution to (5.3) if and only if there exists $z \in \mathbb{R}^n$ such that*

$$w = X^+ Y + (\mathcal{I}_n - X^+ X)z.$$

Proof. Firstly, note that for any $z \in \mathbb{R}^n$, if $w = X^+ Y + (\mathcal{I}_n - X^+ X)z$, then

$$Xw = XX^+ Y + (X - XX^+ X)z = XX^+ Y + (X - X)z = XX^+ Y,$$

where we used item 1 in Definition 5.1.1. We see that whatever value z takes, it vanishes when X is applied to w. Therefore let $w' := X^+ Y' \in \mathrm{Ker}(X)^\perp$ for an arbitrary $Y' \in \mathbb{R}^m$. We now check that if $w' \neq X^+ Y$, then $C(w) < C(w')$. We write

$$C(w') = \|Xw' - Y\|^2 = \|Xw - Y + X(w' - w)\|^2.$$

Note that $Xw - Y = (XX^+ - \mathcal{I}_m)Y \in \mathrm{Im}(X)^\perp$ by Lemma 5.1.2. Moreover, $X(w' - w) \in \mathrm{Im}(X)$, so that $X(w' - w) \perp Xw - Y$. The Pythagorean theorem thus shows that

$$\|Xw - Y + X(w' - w)\|^2 = \|Xw - Y\|^2 + \|X(w' - w)\|^2 = C(w) + \|X(w' - w)\|^2.$$

We see that whenever $X(w' - w)$ is nonnull, $C(w') > C(w)$. By Lemma 5.1.2 and the discussion below its proof we have that $X(w' - w) = XX^+(Y' - Y)$ is null if and only if $Y' - Y \in \mathrm{Im}(X)^\perp$, that is, if and only if $X^+ Y' = X^+ Y$. This shows that $w = X^+ Y + (\mathcal{I}_n - X^+ X)z$ are the only solutions of (5.3) and concludes the proof. $\square$

Corollary 5.1.4. *Following Theorem 5.1.3 and assuming that the data $\{x_1, \ldots, x_m\}$ are not colinear, we can specify some properties of the solution w:*
(i)   *When $n = m$, we have by definition $X^+ = X^{-1}$ and thus $w = X^{-1}Y$.*
(ii) *When $m > n$, $X^T X$ is invertible, and there is a unique $w = (X^T X)^{-1} X^T Y$.*
(iii) *When $n > m$, $X^T X$ is not invertible, and there are infinitely many solutions w.*

The proof of this corollary is left as an exercise to the reader.

In case (iii) of Corollary 5.1.4, we find an infinite number of w^* that achieve the same minimal squared error. So we may ask which solution do we seek in this case (i. e., when $X^T X$ is not invertible)? We can seek to solve the following minimization problem to obtain a unique solution w instead:

$$w^* = \arg\min_w \|w\|^2 \quad \text{subject to} \quad \min_w \|Xw - y\|^2, \tag{5.4}$$

that is, we find a solution with minimal L_2 norm *in the parameter space*. In this case, we have the following result.

Corollary 5.1.5. *The solution to* (5.4) *is given by* $w = X^+Y$.

The result follows from Theorem 5.1.3, by setting $z = \vec{0}$, as it does not change the cost and minimizes the norm of w.

Remark 5.1.6 (Over- and underdetermined systems). We discussed two cases for the linear regression problem: one where we have an invertible matrix $(X^TX)^{-1}$ and another where we do not. The discussion is also often made with respect to having an *overdetermined* system when $m > n$, which often leads to an invertible $(X^TX)^{-1}$ corresponding to case (ii), or *underdetermined* system when we have more degrees of freedom than data to constrain the system, i. e., $n > m$, which corresponds to case (iii) of Corollary 5.1.4.

Remark 5.1.7 (Direction of a constructive proof of Corollary 5.1.4(ii)). When X^TX is invertible, then we can directly derive the form of w by evaluating the gradient of objective (5.3) and setting it to zero. Then by showing that the Hessian of the objective is semipositive definite we see that the minimization is convex, so that the w found is the unique minimum for the optimization problem. This is sketched later on for the ridge regression case.

Remark 5.1.8 (No closed-form solution). In generality, we might not have access to the closed-form solution of a learning task. Then we can use optimization techniques, such as **gradient descent**, as introduced in Chapter 3.

But even when a closed-form solution is available such as for linear models, we might use gradient descent. This is because building $(X^TX)^{-1}$ can be expensive, whereas computing the gradient of C and iterating is cheap. Furthermore, because C is convex, we find the optimal solution (for appropriate learning rate).

5.1.3 Regularization

In (5.4), we modified the objective to include some norm of w to achieve a unique solution. Indeed, (infinitely) many solutions may have zero empirical loss, but we hope to have a small generalization error as well. The practice of adding something extra to the minimization is a way to prevent overfitting and attain a unique solution.

Regularized Loss Minimization (RLM) is a learning rule in which we jointly minimize the empirical risk (ERM) and a regularization function.

A regularization function is a mapping $\mathcal{R} : \mathbb{R}^P \to \mathbb{R}_+$, and the regularized loss minimization rule outputs a hypothesis in

$$\arg\min_{w}(L(w) + \mathcal{R}(w)).$$

Suppose that the regularization function measures the complexity of hypotheses. Then the above minimizes the sum of the empirical risk and penalization on the complexity of the model. Hence this prevents overfitting by favoring simpler hypotheses. The complexity of w can be seen by its norm, so we can take the L_p norm to regularize w.

The L_2 regularization is standard and is given by

$$R(w) = \lambda\|w\|^2, \quad \lambda > 0,$$

where $\|w\|^2 = \sum_{j=1}^{n} w_j^2$, which in the context of linear regression leads to the **Ridge regression** defined by the following minimization problem:

$$\min_w \frac{1}{2m} \sum_{i=1}^{m} (y_i - w^T x_i)^2 + \lambda\|w\|^2, \quad \lambda > 0. \tag{5.5}$$

Theorem 5.1.9. *Consider the ridge regression problem (5.5). If $-m\lambda$ is not an eigenvalue of $X^T X$, then the solution is unique and is given by*

$$w^* = (X^T X + \lambda m I_m)^{-1} X^T Y,$$

where I_m is the identity matrix in $\mathbb{R}^{m \times m}$.

Proof. We can show that the map

$$L(w) + R(w)$$

is convex. To find all the solutions, we take the gradient with respect to w and get

$$\begin{aligned}
\nabla_w(L(w) + R(w)) &= \frac{1}{m} \nabla_w(\|Y\|^2 - 2w^T X^T Y + w^T X^T X w + m\lambda\|w\|^2) \\
&= \frac{1}{m}(-2X^T Y + 2X^T X w + 2m\lambda w) \\
&= \frac{1}{m}(2(X^T X + m\lambda I_m)w - 2X^T Y).
\end{aligned}$$

The above gradient is zero if and only if $w = (X^T X + m\lambda I_m)^{-1} X^T Y$, which is therefore the unique solution of (5.5), as claimed, which concludes the proof. $\square$

Another commonly used norm for regularization is the L_1 norm, namely

$$R(w) = \lambda\|w\|_1, \quad \|w\|_1 = \sum_{i=1}^{n} |w_i|,$$

which defines the **Lasso regression**, leading to the optimization problem

$$\min_w \frac{1}{2m} \sum_{i=1}^{m} (y_i - w^T x_i)^2 + \lambda\|w\|_1, \quad \lambda > 0.$$

However, there is no closed-form solution for the general case.

Remark 5.1.10. Note that a minimal norm solution (as seen in Corollary 5.1.5) **is not** a solution of the norm-regularized problem. Nevertheless, it can be shown that as $\lambda \to 0$, the solution of the L_2 regularized problem converges to the minimal L_2 norm solution of the original (nonregularized) problem.

The example below compares minimal L_1 and L_2 norm solutions, as well as L_2 regularized solutions.

Example 5.1.11. Let $x_1 = \begin{pmatrix} 1 \\ 1 \end{pmatrix}$, $x_2 = \begin{pmatrix} 2 \\ 2 \end{pmatrix}$, $y_1 = 1$, and $y_2 = 2$. We can see that the solutions of $\min_{w \in \mathbb{R}^2} \sum_{i=1}^{2}(w^T x_i - y_i)^2$ are given by the line $w = \begin{pmatrix} a \\ 1-a \end{pmatrix}$, $a \in \mathbb{R}$.

The minimal L_2 norm solution is obtained for $a = 1/2$, whereas the segment $a \in [0,1]$ contains all minimal L_1 norm solutions of the problem.

However, if we regularize with an L_2 penalty on the parameters, i. e., we minimize $\min_{w \in \mathbb{R}^2} \frac{1}{4} \sum_{i=1}^{2}(w^T x_i - y_i)^2 + \lambda \|w\|_2^2$ for some $\lambda > 0$, then $w = \begin{pmatrix} 1/2 \\ 1/2 \end{pmatrix}$ is not a solution (we can check this by taking the gradient and noting that it is nonzero).

The conclusion is similar for the L_1-regularized problem.[1]

Remark 5.1.12. Note that now the scale of the vectors in X matters. Why? We want $\vec{w}$ to be small; if features (i. e., coordinates) span different magnitudes, then the contribution of a large feature will dominate the regression. So it is necessary, when considering Ridge or Lasso regression, to center and normalize the data X.

5.1.4 Representing nonlinear functions using basis functions

As we have seen, linear models have nice theoretical guarantees. What if the relationship between inputs and outputs is not linear? It turns out that we can use linear models to express nonlinear relationships. Indeed, suppose we have data given in the form similar to Figure 5.2. We can see that our predictor can benefit from having nonlinear features. Namely,

$$f_w(x) = w_1 x^2 + w_0 x + b.$$

Our linear coefficients are $w = [b, w_0, w_1]$, and our features become $[1, x, x^2]$. We are finding a linear model on *nonlinear features*, namely, given by x^k, $k = 0, 1, 2$.

In general, we can fit nonlinear functions via linear regression using a transformation ϕ that applies nonlinear transformations on our features:

$$f_w(x) = \sum_{i=1}^{d} w_i \phi_i(x).$$

---

1 An example of regression weights optimized under different regularization strategies: https://developers.google.com/machine-learning/crash-course/regularization-for-sparsity/l1-regularization

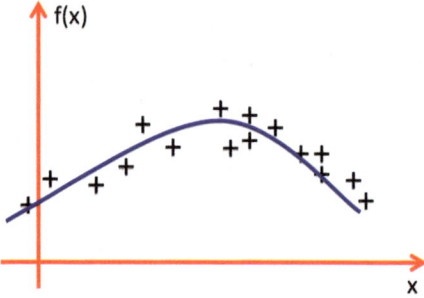

Figure 5.2: Quadratic relation.

For example, if we consider polynomial transformations of our feature space, then ϕ can be given as

- Feature space is one-dimensional: $\phi(x) \rightarrow (1, x, x^2, \ldots, x^k)$.
- Feature space is two-dimensional: $\phi(x) \rightarrow (1, x_1, x_2, x_1^2, x_2^2, x_1 x_2, \ldots, x_1^k, x_2^k)$.
- Feature space is d-dimensional: a vector of all monomials in x_1 to x_p of degree up to k.

5.2 Classification

Let us turn again to the classification problem, where we consider a dataset with binary labels, i. e., $S = \{(x_i, y_i) : i = 1, \ldots, m\} \subset \mathbb{R}^n \times \{-1, 1\}$.

Assumption. We suppose that the dataset $S \subset \mathbb{R}^n \times \{-1, 1\}$ is *linearly separable*, i. e., there exists a hyperplane $P = \{x \in \mathbb{R}^n : w^T x + b = 0\}$ for some fixed $w \in \mathbb{R}^n_*$ and $b \in \mathbb{R}$ that separates the data. Formally, for all $i = 1, \ldots, m$, $y_i = 1$ if and only if $w^T x_i + b > 0$, and, similarly, $y_i = -1$ if and only if $w^T x_i + b < 0$.

The class of half-spaces is defined as follows:

$$\mathcal{H} = \{f_w : x \rightarrow f_w(x) = \text{sign}(\vec{w}^T \vec{x} + b) : \vec{w} \in \mathbb{R}^n, b \in \mathbb{R}\}.$$

Let us illustrate this hypothesis class geometrically, considering the case $n = 2$. Each hypothesis forms a hyperplane (for $n = 2$, a line) that is orthogonal to the vector $\vec{w} = (w_1, w_2)$ and intersects the vertical axis at the point $(0, -b/w_2)$. The instances that are "above" the hyperplane, that is, share an acute angle with $\vec{w}$, are labeled positively. Instances that are "below" the hyperplane, that is, share an obtuse angle with $\vec{w}$, are labeled negatively.

How do we find a good $\vec{w}$? We can consider the following minimization:

$$\arg\min_{w} C(w) = \arg\min_{\vec{w}, b} \sum_{i=1}^{m} \mathbb{1}_{\{y_i \neq \text{sign}(\vec{w}^T \vec{x}_i + b)\}}.$$

However, the loss function $\ell(y, y') = \mathbb{1}_{\{y \neq y'\}}$, called **0–1 loss**, is neither convex nor continuous. Thus, to make the optimization problem easier, we introduce a surrogate loss, called the **perceptron loss**:

$$\ell_{\text{perc}}(y, y') = \max(0, -yy'),$$

and the cost function becomes[2] $C(w) = \frac{1}{m} \sum_{i=1}^{m} \max(0, -y(\vec{w}^T \vec{x} + b))$.

5.2.1 Perceptron algorithm

The perceptron loss is useful for solving the optimization problem.[3] As previously seen, we compute the gradient of the cost:

$$C(w) = \frac{1}{m} \sum_{i=1}^{m} \max(0, -y_i(w^T x_i + b)),$$

$$\nabla C(w) = \frac{1}{m} \sum_{i=1}^{m} \nabla_w \max(0, -y_i(w^T x_i + b))$$

$$= \frac{1}{m} \sum_{i=1}^{m} \begin{cases} 0 & \text{if } y_i w^T x_i + b \geq 0 \text{ (correctly classified)}, \\ -y_i x_i & \text{otherwise (misclassified)}. \end{cases}$$

Then we can use the gradient descent algorithm defined in Definition 3.2.1 to find a plane that linearly separates the data (if possible). This gives us the *perceptron algorithm*

$$w_{t+1} = w_t - \eta_t \frac{1}{m} \sum_{i: y_i(w^T x_i + b) < 0} -y_i x_i.$$

The perceptron algorithm guarantees that the trained predictor separates the data if the data is linearly separable (i. e., if gradient descent converges, then it finds a local minimum, which by convexity is a global minimum). There are in general, however, infinitely many planes that separate the data, and we do not know a priori which one will be found.

5.2.2 Support vector machine

The support vector machine (SVM) is a linear classifier that can be viewed as an extension of the perceptron algorithm. In the context of binary classification in a linearly sepa-

---

2 Note that if y and $w^T x - b$ have the same sign, then the second term is negative, and thus $\ell_{\text{perc}}(x, y)$ is zero. If the signs are opposite, then the error will be positive.

3 What about at $x = 0$? The derivative is not unique, but it can be approached with subgradients.

rable dataset, the perceptron guarantees that you find a separating hyperplane, whereas the SVM finds the **maximum-margin separating hyperplane**. Refer to Figure 5.3 for a comparison.

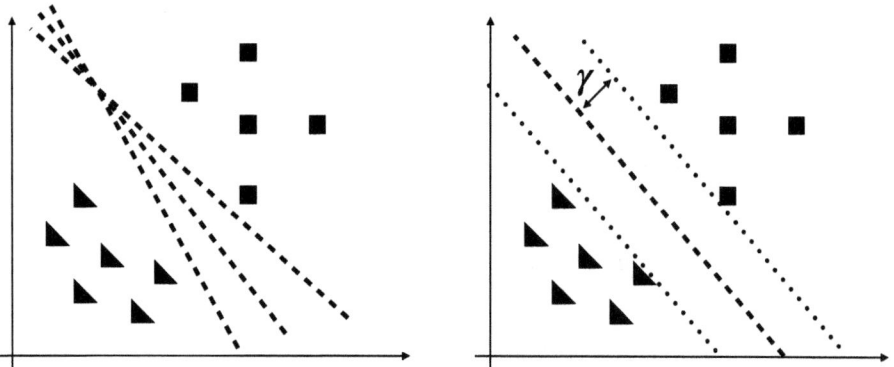

Figure 5.3: Two different separating hyperplanes for the same data set. Right: The maximum margin hyperplane. The margin y is the distance from the hyperplane (solid line) to the closest points in either class (which touch the parallel dotted lines).

Definition 5.2.1 (Margin). Consider a separating hyperplane defined by w and b as the set $P = \{x \in \mathbb{R}^n : w^T x + b = 0\}$. The margin $y(w, b)$ is defined as the distance from the hyperplane to the closest point across both classes.

Given a linearly separable dataset $S = \{(x_i, y_i); i = 1, \ldots, m\}$, we can state the **SVM objective**

$$\max_{w,b} y(w, b) \quad \text{s. t.} \quad \forall i = 1, \ldots, m, \quad y_i(w^T x_i + b) \geq 0. \tag{5.6}$$

By maximizing the SVM objective we seek the hyperplane defined by w and b such that the distance of that plane to points in the dataset is maximized and such that the plane correctly classifies points.

The expression above is not yet computable. Namely, we need an explicit expression for the margin y to solve this optimization problem. However, we will see that it leads us to the max-min problem (5.6), which is not trivial to solve. Luckily, it turns out that the SVM objective can be reformulated in a much *nicer* way.

Theorem 5.2.2. *The solution (w_*, b_*) of*

$$\max_{w,b} y(w, b) \quad \text{s. t.} \quad \forall i = 1, \ldots, m, \quad y_i(w^T x_i + b) \geq 0$$

is equal to the solution (w'_*, b'_*) of

$$\min_{w} w^T w$$

$$s.t. \quad \forall i = 1, \ldots, m, \quad y_i(w^T x_i + b) \geq 1. \tag{5.7}$$

Remark 5.2.3. This new formulation (5.7) is a quadratic optimization problem. The objective is quadratic, and the constraints are all linear. We can solve it efficiently with a Quadratically Constrained Quadratic Program solver (e.g., see [2]). It has a unique solution whenever a separating hyper plane exists.

Before we prove Theorem 5.2.2, we need to establish some preliminary results such as finding a convenient expression for the margin of a separating hyperplane. Namely, how do we find the margin or how do we compute the distance of an arbitrary point x to a hyperplane P (see Figure 5.4)?

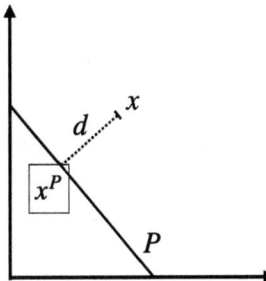

Figure 5.4: Distance between an arbitrary point x and hyperplane P with the projection of x onto P as x^P.

Lemma 5.2.4. Let P be the hyperplane induced by a nonzero $w \in \mathbb{R}^n$ and $b \in \mathbb{R}$, and let $x \in \mathbb{R}^n$. We have that

$$\text{dist}(P, x) = \frac{|w^T x + b|}{\|w\|}.$$

In particular, if $S = \{(x_1, y_1), \ldots, (x_m, y_m)\}$, then the margin of P with respect to S is given by

$$\gamma(w, b) = \min_{x \in S} \frac{|w^T x + b|}{\|w\|}.$$

Furthermore, we can rescale the parameters w and b, and thus the margin, without changing the hyperplane (i.e., invariant to rescaling)

$$\gamma(\beta w, \beta b) = \gamma(w, b) \quad \forall \beta \in \mathbb{R}, \beta \neq 0.$$

Proof. Let $d := x - x^P \in \mathbb{R}^n$, where x^P is the orthogonal projection of x onto hyperplane P. We know that $w^T x^P + b = 0$ by definition, which entails that

$$w^T(x - d) + b = 0.$$

What is d? It is the distance vector resulting from subtracting x from its projection x^P, and its norm is the minimum distance between x and any point on the hyperplane. Furthermore, note that d is colinear with w, so we can write $d = aw$ for some $a \in \mathbb{R}$. Then

$$w^T(x - d) + b = 0,$$
$$w^T(x - aw) + b = 0,$$
$$a = \frac{w^T x + b}{w^T w} = \frac{w^T x + b}{\|w\|^2}.$$

Since now we know a, we can compute the length of $d = aw$, i. e., the distance of x to P as

$$\|d\| = \sqrt{d^T d} = \sqrt{a^2 w^T w} = \frac{|w^T x + b|}{\|w\|},$$

as claimed.

The margin of the hyperplane P to a dataset S follows by definition, i. e., we find the point in S for which the distance to the hyperplane is the smallest. From the expression of the margin we then deduce its scale-invariance property, which concludes the proof. □

Remark 5.2.5 (Symmetry of the margin). If the hyperplane is such that the margin γ to a labeled dataset is maximized, then it must lie right in the middle of the two classes. In other words, γ must be the distance to the closest point within both classes. (If not, then you could move the hyperplane toward data points of the class that is further away and increase γ, which contradicts that γ is maximized.)

Proof of Theorem 5.2.2. By plugging the expression of γ from Lemma 5.2.4 into objective (5.6) we get

$$\max_{w,b} \left[\min_{x \in S} \frac{|w^T x + b|}{\|w\|} \right] \quad \text{such that } \forall i \in \{1, \dots, m\}, \; y_i(w^T x_i + b) \geq 0.$$

We can pull the denominator outside of the minimization because it does not depend on x. Because the hyperplane is scale invariant, we can fix the scale of w and b any way we want. Let us be clever about it, and choose it such that

$$\min_{x \in S} |w^T x + b| = 1.$$

Then we can simplify the optimization:

$$\max_{w} \frac{1}{\|w\|} \cdot 1 = \min_{w} \|w\|;$$

note that the w that (satisfies the constraints and) minimizes $\|w\|$ is the same that minimizes $\|w\|^2$. This is because $f(x) = x^2$ is increasing for $x \geq 0$ and $\|w\| \geq 0$.

Then the new constrained optimization problem becomes

$$\min_w w^T w$$

$$\forall i \in \{1, \ldots, m\}, \quad y_i(w^T x_i + b) \geq 0$$

$$\text{s.t.} \quad \min_i |w^T x_i + b| = 1.$$

These constraints are still hard to deal with. However, luckily, we can show that (for the optimal solution) they are equivalent to the much simpler (5.7).

($\Longrightarrow$) We can write the constraints with the absolute value

$$|w^T x_i + b| = 1$$

as

$$y_i(w^T x_i + b) = 1,$$

because we are in the separable case. If the minimum is 1, then all other points are ≥ 1.

($\Longleftarrow$) Assume that for all $i \in \{1, \ldots, m\}, y_i(w^T x_i + b) \geq 1$ (e. g., larger than 1). Then how can we guarantee that $|w^T x_i + b| = 1$ for at least one point x_i? Suppose it is not the case. Then w is not minimized, and we can minimize it further. (Divide w and b by $\|w\|$.) This concludes the proof. $\square$

Now that we have established the simpler formulation of the SVM problem (5.7), we can numerically find a solution to it as mentioned in Remark 5.2.3. Although we do not have a general closed-form solution, more can be said about the maximal margin hyperplane by studying the optimization problem derived in Theorem 5.2.2, which is a constrained optimization problem.

For a constrained optimization problem, we can derive its Lagrangian. The Lagrangian for the SVM problem is

$$\mathcal{L}(w, b, \beta) = w^T w - \sum_{i=1}^{m} \beta_i(y_i(w^T x_i + b) - 1) \tag{5.8}$$

for $\beta \in [0, \infty)^m$. The solution (w^*, b^*, β^*) to the minimax problem

$$\min_{w,b} \max_{\beta} \mathcal{L}(w, b, \beta)$$

yields the solution (w^*, b^*) to (5.6). This means that we could, in principle, solve an unconstrained optimization with the Lagrangian formulation with a minimax optimization instead of a constrained optimization as in (5.6).

Furthermore, the solution (w^*, b^*, β^*) above is also the solution to the maximin problem

$$\max_{\beta} \min_{w,b} \mathcal{L}(w, b, \beta).$$

This is only true because the SVM optimization problem satisfies the strong duality condition.[4]

We recall Theorem 3.4.3, which tell us that the optimal solution (w^*, b^*, β^*) satisfies the Karush–Kuhn–Tucker (KKT) conditions. We will use the KKT conditions to say something more about the structure of the optimal solution to the SVM problem.

The stationarity condition:

$$\nabla_w \mathcal{L} = 2w^* - \sum_{i=1}^{m} \beta_i^* y_i x_i = 0 \quad \implies \quad w^* = \sum_{i=1}^{m} \beta_i^* y_i x_i. \tag{5.9}$$

This means that the optimal weight vector w^*, which defines the normal to the hyperplane that maximizes the margin, is a linear combination of the training data $\{x_1, \ldots, x_m\}$.

The complementary slackness condition:

$$\forall i \in \{1, \ldots, m\}, \quad \beta_i^* [y_i(w^T x_i + b) - 1] = 0 \quad \implies \quad \beta_i^* = 0 \text{ or } y_i(w^T x_i + b) = 1. \tag{5.10}$$

This means that there are some training datapoints that lie on the margin, and for those, the corresponding β_i^* is nonzero.

Combining (5.9) and (5.10), a training datapoint x_i appears in the expansion of w^* if and only if $\beta_i^* \neq 0$. The training datapoints that appear in the expansion of w^* are called *"support vectors"*. The support vectors define the maximum margin hyperplane.[5]

Remark 5.2.6 (Nonuniqueness of support vectors). Although w is unique for the SVM problem, the support vectors are not. For example, in a hyperplane in n dimensions, we need $n + 1$ points to define a hyperplane. If there are more than $n + 1$ support vectors, then we can choose different support vectors to specify the same hyperplane.

Remark 5.2.7 (Moving training datapoints). Note that if we move one of the support vectors and retrain the SVM, then the resulting hyperplane would change. However, if we move nonsupport vectors, then the SVM hyperplane would not change (provided that they do not move too much or they could turn into support vectors themselves).

---

4 For a reminder of essential results in constrained optimization, refer to Section 3.4.

5 To find b, note that

$$y_j(w^T x_j + b) = 1$$

for some j (one of the support vectors). Then $b = y_j - w^T x_j$.

We can wrap up the discussion above in the following:

Theorem 5.2.8. *The SVM classifier f_* with parameters (w^*, b^*) that maximize the maximal margin hyperplane for a separable dataset $S = \{(x_i, y_i); i = 1, \ldots, m\}$ is of the form*

$$f_*(x) = \text{sign}((w^*)^T x + b^*) = \text{sign}\left(\left(\sum_{i=1}^{m} \beta_i^* y_i x_i \right)^T x + b^* \right)$$

for $\beta_i \in [0, \infty)$.

Note that the approximating function is given by an inner product between support vectors and new datapoint x, which we will exploit in the next chapter to learn nonlinear hyperplanes.

5.2.3 Nonseparable case

So far, we have assumed that the data were linearly separable. This is often not the case; it is possible that there is no hyperplane that separates the two classes. In this case, there is no solution to the optimization problems stated previously.

We can fix this by allowing the constraints to be violated ever so slightly with the introduction of slack variables $\xi = (\xi_1, \ldots, \xi_m)$:

$$\min_{w, b, \xi} w^T w + \lambda \sum_{i=1}^{m} \xi_i$$
$$\text{s. t.} \quad \forall i \in \{1, \ldots, m\}, \; y_i(w^T x_i + b) \geq 1 - \xi_i, \quad \lambda \geq 0, \tag{5.11}$$
$$\xi_i \geq 0.$$

The slack variable ξ_i allows the input x_i to be closer to the hyperplane (or even be on the wrong side), but there is a penalty in the objective function for such a "slack", given by λ:

- If λ is very large, then the SVM becomes very strict and tries to get all points to be on the right side of the hyperplane.
- If λ is very small, then the SVM becomes very loose and may "sacrifice" some points to obtain a *simpler* hyperplane (i. e., lower $\|w\|_2^2$) solution.

We can write (5.11) as an unconstrained optimization by noting that for $\lambda \neq 0$, we have

$$\xi_i = \begin{cases} 1 - y_i(w^T x_i + b) & \text{if } y_i(w^T x_i + b) < 1, \\ 0 & \text{if } y_i(w^T x_i + b) \geq 1. \end{cases}$$

This is equivalent to the following closed form:

$$\xi_i = \max(1 - y_i(w^T x_i + b), 0).$$

If we plug this closed form into the objective of our SVM optimization problem (5.11), then we obtain the following unconstrained optimization composed of two terms:

$$\min_{w,b} \underbrace{w^T w}_{l_2\text{-regularizer}} + \lambda \sum_{i=1}^{m} \underbrace{\max[1 - y_i(w^T x_i + b), 0]}_{\text{hinge-loss}}.$$

5.3 Implementation details

In this section, we provide some code snippets for linear regression and SVM using the scikit-learn library.

Listing 5.1: Linear regression model using scikit-learn

```
 1 import numpy as np
 2 from sklearn.model_selection import train_test_split
 3 from sklearn import datasets, linear_model
 4 from sklearn.metrics import mean_squared_error
 5
 6 # Load generic dataset for regression
 7 X, y = datasets.load_diabetes(return_X_y=True)
 8
 9 # Create hold-out test set
10 X_train, X_test, y_train, y_test = train_test_split(X, y,
       test_size=0.2, random_state=0)
11
12 regressor = linear_model.LinearRegression()
13 regressor.fit(X_train, y_train)
14
15 # Evaluate performance on test set
16 y_pred = regressor.predict(X_test)
17 print(f"MSE: {mean_squared_error(y_test, y_pred)}")
```

Listing 5.2: SVM model using scikit-learn

```
 1 from sklearn.datasets import load_breast_cancer
 2 from sklearn.model_selection import train_test_split
 3 from sklearn import svm
 4
 5 # Standard binary classification task
 6 X, y = load_breast_cancer(return_X_y=True)
 7 X_train, X_test, y_train, y_test = train_test_split(X, y)
```

```
 8
 9 classifier = svm.LinearSVC()
10 classifier.fit(X_train, y_train)
11
12 print(f"Mean accuracy: {classifier.score(X_test,y_test)}")
```

Additional coding resources that were featured in this chapter can be found in the github: https://github.com/hanveiga/tmml:
–   Optimizing with L1 or L2 losses.
–   Implementation of perceptron.

6 Kernel methods

Linear classifiers are great, but what if the decision boundary is highly nonlinear? As it turns out, there is an elegant way to incorporate nonlinearities into most linear classifiers.

We can make linear classifiers nonlinear by applying a nonlinear mapping ϕ on the input feature vectors $\mathcal{X}$ to a higher-dimensional space $\mathcal{X}_H$, where linear separation is possible; see the example in Figure 6.1. However, there are some disadvantages to this approach; for example, $\phi(x)$ may be very high dimensional, and building $\phi(x)$ from scratch may be difficult.

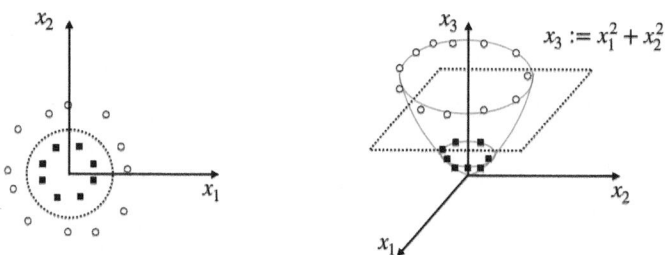

Figure 6.1: Nonlinear transformation on the input. On the left, the correct boundary decision in two dimensions is a circle, so that linear decision boundaries cannot separate the data. Transforming the data by adding the third coordinate $x_3 = x_1^2 + x_2^2$ to the data, we get the figure on the right, where a linear decision boundary – a plane – perfectly separates the data.

In this chapter, we talk about methods that use this idea of sending elements of $x \in \mathcal{X}$ onto some higher-dimensional space $\mathcal{X}_H$, which we do not necessarily have to know much about, where our problem becomes easier to solve. All we need to know about this new space is that it is a so-called *reproducible kernel Hilbert space.*

Before we can dive into what this means, we start with a brief review and introduction to key concepts.

Note: throughout this chapter, to ease the notation, we always assume that the output space $\mathcal{Y} = \mathbb{R}$.

6.1 Inner products and kernels

In $\mathbb{R}^n$ the dot product $x \cdot x' = \sum_{i=1}^{n} x_i x_i'$ can be seen as a way to measure the similarity between two elements x, x': $x \cdot x' = 0$ if and only if x and x' are orthogonal, i. e., nothing of x can be used to represent x', and $x \cdot x' \leq \|x\| \cdot \|x'\|$ with equality if and only if $x = x'$, that is, the dot product is maximal if and only if $x = x'$. In an arbitrary vector space, this notion can be generalized:

https://doi.org/10.1515/9783111288994-006

Definition 6.1.1. An *inner product* in a real vector space H is a map $\langle \cdot, \cdot \rangle : H \times H \to \mathbb{R}$ that satisfies the following properties: for all $u, v, w \in H$ and $\alpha \in \mathbb{R}$,

1. $\langle u, v \rangle = \langle v, u \rangle$ (symmetry).
2. $\langle u + \alpha v, w \rangle = \langle u, w \rangle + \alpha \langle v, w \rangle$ (bilinearity).
3. $\langle v, v \rangle \geq 0$ with equality if and only if $v = 0$ (strict positive-definiteness).

Note that by symmetry, bilinearity is equivalent to linearity in the first variable. To avoid ambiguity, we will sometimes write $\langle \cdot, \cdot \rangle_H$ to specify on which space we consider the inner product.

Example 6.1.2 (Inner product examples). Some common spaces and associated inner products are:

- The Euclidean space $\mathbb{R}^n$, where the inner product is given by the dot product

$$\langle (x_1, \ldots, x_n), (y_1, \ldots, y_n) \rangle = x_1 y_1 + \cdots + x_n y_n.$$

- The vector space of square-integrable real functions on a closed interval $[a, b]$ with inner product

$$\langle f, g \rangle = \int_a^b f(x)g(x)dx.$$

This inner product induces the norm on H defined by

$$\|u\|_H := \sqrt{\langle u, u \rangle_H}.$$

Definition 6.1.3 (Bounded linear operator). Let $(H, \|\cdot\|_H)$ be a normed space. A map $T : H \to \mathbb{R}$ is called a *linear operator* if

$$T(u + \lambda v) = T(u) + \lambda T(v)$$

for all $u, v \in H$ and $\lambda \in \mathbb{R}$.

We say that T is a *bounded operator* if

$$\exists M > 0 \quad \text{s.t.} \quad \forall u \in H, \ |T(u)| \leq M \|u\|_H.$$

Remark 6.1.4. The above notion of boundedness differs from the usual one in that when we say that a map $f : \mathbb{R} \to \mathbb{R}$ is bounded, we mean that there exists $C > 0$ such that $|f(x)| \leq C$ for all $x \in \mathbb{R}$. On the other hand, a nonzero linear operator T on a normed space is never "bounded" in the standard sense since $T(\lambda u) = \lambda T(u)$, which diverges as $\lambda \to \infty$ whenever $T(u) \neq 0$; it is only "bounded" in the standard sense on the closed unit ball $\{u \in H; \|u\|_H \leq 1\}$.

Sometimes, bounded linear operators are instead called *continuous linear opera-tors*, since for a normed space, a linear operator is bounded if and only if it is Lipschitz continuous, which can be shown using the linearity.

Example 6.1.5. Consider the normed space ℓ^2 of real sequences $(x_n)_{n\geq1}$ such that $\sum_{n\geq1} x_n^2 < \infty$. Then the shift operator $T((x_n)_{n\geq1}) := (x_{n+1})_{n\geq1}$ is bounded, but the operator $I((x_n)_{n\geq1}) := (nx_n)_{n\geq1}$ is not bounded. (Both T and I are linear.)

In Definition 6.1.3 the smallest M such that $|T(u)| \leq M\|u\|_H$ for all $u \in H$ is called the *operator norm* of T and is denoted by $\|T\|_{\mathrm{op}}$.

From an inner product we defined a norm, and from a norm we can define the metric $d(u, v) := \|u - v\|_H$, so that the notion of *completeness* of a space makes sense.

Definition 6.1.6. We say that a sequence $(u_n)_{n\geq1}$ in an inner product space H is *a Cauchy sequence* if for all $\epsilon > 0$, there exists $N \geq 1$ such that for all $m, n \geq N$, it holds that

$$\|u_n - u_m\|_H < \epsilon.$$

An inner product space H is said to be *complete* (with respect to norm $\|\cdot\|_H$) if any Cauchy sequence $(u_n)_{n\geq1}$ of elements of H converges with respect to norm $\|\cdot\|_H$ to some $u_* \in H$.

An inner product space that is complete with respect to the norm induced by its inner product is called a *Hilbert space*.

Theorem 6.1.7 (Riesz representation theorem). *If T is a linear bounded operator on a Hilbert space H, then there exists a unique $g \in H$ such that*

$$T(f) = \langle f, g \rangle_H \quad \forall f \in H.$$

This means that every bounded linear operator $T : H \to \mathbb{R}$ has a unique represen-tative $g \in H$ such that T corresponds to the inner product with g.

Definition 6.1.8. Let $\mathcal{X}$ be an nonempty set. We say that a symmetric[1] function $K : \mathcal{X} \times \mathcal{X} \to \mathbb{R}$ is a *positive definite kernel* (pd kernel) if for any fixed $n \in \mathbb{N}$ and $c_1, \ldots, c_n \in \mathbb{R}$, we have

$$\sum_{i,j=1}^{n} c_i c_j K(x_i, x_j) \geq 0 \quad \forall x_1, \ldots, x_n \in \mathcal{X}.$$

Remark 6.1.9. In some fields such as probability theory, the above is called a *positive semidefinite kernel*, because it allows the sum in the definition to be 0 even for nonnull c_i. In the machine learning literature, a "positive definite kernel" is understood as in Def-inition 6.1.8. This is the convention we will use in this book, and when the inequality is always strict for nonnull c_is, we call the kernel *strictly* positive definite.

---

1 In this book, a kernel is always symmetric.

Note that there are no restrictions on $\mathcal{X}$ (except that it is nonempty). The property that defines pd kernels can be rephrased as follows: for all $n \in \mathbb{N}$ and $x_1, \ldots, x_n \in \mathcal{X}$, the matrix $\underline{K}$ given by

$$\underline{K}_{i,j} = K(x_i, x_j), \quad i,j = 1, \ldots, n,$$

is (symmetric) positive semidefinite. This matrix is called the *Gram matrix*.

Exercise. Show that it is equivalent for a symmetric matrix to be positive semidefinite and to have all its eigenvalues nonnegative.

6.2 Reproducible kernel Hilbert spaces

We introduced two notions of inner products and pd kernels in the previous section. The link between them may not seem straightforward. However, as mentioned in the previous section, pd kernels share some similarities with pd symmetric matrices, which satisfy the following: if $A \in \mathbb{R}^{n \times n}$ is a strictly pd symmetric matrix, then the function $b : (x, x') \mapsto x^T A^{-1} x', x, x' \in \mathbb{R}^n$, defines an inner product.

Viewing pd kernels as an infinite-dimensional analogue of pd symmetric matrices, we can hope that a pd kernel is always linked to some inner product. It is indeed the case, but before we see it, we need to introduce some concepts.

Let $\mathcal{H}$ be a Hilbert space of functions from $\mathcal{X}$ to $\mathbb{R}$.

Remark 6.2.1. Suppose that for all $x \in \mathcal{X}$, the functional $L_x : \mathcal{H} \to \mathbb{R}$ defined by $L_x(f) = f(x)$ is a bounded operator on $\mathcal{H}$, i. e.,

$$\forall x \in \mathcal{X}, \quad \exists M_x > 0 \quad \text{s.t.} \quad \forall f \in \mathcal{H}, \quad |f(x)| \leq M_x \|f\|_{\mathcal{H}}. \tag{6.1}$$

Then by the Riesz representation theorem there exists a unique $k_x \in \mathcal{H}$ such that $L_x(f) = \langle f, k_x \rangle_{\mathcal{H}}$, so that we can define the kernel $k : (x, x') := \langle k_x, k_{x'} \rangle_{\mathcal{H}}$ on $\mathcal{X} \times \mathcal{X}$ such that $\mathcal{H}$ is an RKHS.

Definition 6.2.2. We say that a kernel K on $\mathcal{H}$ satisfies the *reproducing property* if for all $x \in \mathcal{X}$ and $f \in \mathcal{H}$, we have

$$\langle f, K_x \rangle_{\mathcal{H}} = f(x),$$

where $K_x := K(x, \cdot) \in \mathcal{H}$. In this case, we say that $\mathcal{H}$ is a *reproducible kernel Hilbert space* (RKHS) with reproducing kernel K.

We note in particular that for all $x, y \in \mathcal{X}$,

$$K(x, y) = \langle K_x, K_y \rangle_{\mathcal{H}}. \tag{6.2}$$

Example 6.2.3 (On the reproducing property). Let the feature map $\phi : \mathbb{R}^2 \to \mathcal{H}$, where $\mathcal{H}$ is the space of functions from $\mathbb{R}^2$ to $\mathbb{R}$, be defined for all $x \in \mathbb{R}^2$ by

$$\phi_x \in \mathcal{H} : y \mapsto x_1 y_1 + x_2 y_2 + x_1 x_2 y_1 y_2,$$

$$\mathbb{R}^2 \to \mathbb{R}.$$

Define the kernel K on $\mathbb{R}^2 \times \mathbb{R}^2$ by

$$K(x, x') = \langle \phi_x, \phi_{x'} \rangle_{\mathcal{H}} := x_1 x_1' + x_2 x_2' + x_1 x_2 x_1' x_2'.$$

Fix $u \in \mathbb{R}^3$ and define $f_u : \mathbb{R}^2 \to \mathbb{R}$ by

$$f_u(x) = u_1 x_1 + u_2 x_2 + u_3 x_1 x_2.$$

Note that $f_u \in \mathcal{H}$, since we can write $f_u = \phi_{(u_3,1)} + \phi_{(u_1-u_3,0)} + \phi_{(0,u_2-1)}$, which is a linear combination of elements of $\mathcal{H}$. We thus have an RKHS $\mathcal{H}$ with feature map ϕ, and we can check that K enjoys the reproducing property: since $K_x = \phi_x$, we have

$$\langle f_u, K_x \rangle_{\mathcal{H}} = \langle \phi_{(u_3,1)}, \phi_x \rangle + \langle \phi_{(u_1-u_3,0)}, \phi_x \rangle + \langle \phi_{(0,u_2-1)}, \phi_x \rangle$$
$$= f_u(x),$$

as claimed.

Now given a pd kernel K on $\mathcal{X} \times \mathcal{X}$, we can ask ourselves, is there an RKHS of real functions on $\mathcal{X}$ associated with K? The answer is positive and is known as the *Moore–Aronszajn theorem*.

Theorem 6.2.4 (Moore–Aronszajn theorem). *Let $K : \mathcal{X} \times \mathcal{X} \to \mathbb{R}$ be a positive definite kernel. Then there exists a unique RKHS $\mathcal{H} \subset \{f : \mathcal{X} \to \mathbb{R}\}$ with reproducing kernel K. In particular, there exists a mapping $\phi : \mathcal{X} \to \mathcal{H}$ such that for all $x, x' \in \mathcal{X}$,*

$$K(x, x') = \langle \phi(x), \phi(x') \rangle_{\mathcal{H}}.$$

Note that Theorem 6.2.4 shows that a pd kernel induces a *unique* RKHS and vice versa. However, for a given *RKHS*, the feature map ϕ is not unique. Nonetheless, we can always choose $\phi_x = K_x$ thanks to (6.2). The map $x \mapsto K_x$ is thus called the *canonical feature map*.

In view of Theorem 6.2.4, why do we introduce a kernel instead of computing $\phi(x)$ and then the inner product in $\mathcal{H}$? That is because evaluating the kernel is computationally more tractable. Furthermore, we can define a kernel K without explicitly knowing the space $\mathcal{H}$ and a fortiori its inner product. Thanks to the theorem, a pd kernel allows us to measure the similarity of two points $x, x' \in \mathcal{X}$ through the implicit inner product of $\phi(x)$ and $\phi(x')$ in an unknown RKHS. This is sometimes referred to as the *kernel trick* in machine learning: we send the input space to a higher-dimensional space where the elements can better be compared.

Equipped with those theoretical tools, we now look at some examples where introducing a kernel can be useful.

Example 6.2.5 (Polynomial kernel). For a constant $c > 0$, a polynomial kernel of degree $d \in \mathbb{R}$ is the kernel K defined over $\mathcal{X} \subset \mathbb{R}^N$ by

$$\forall x, x' \in \mathcal{X}, \quad K(x, x') = (x \cdot x' + c)^d.$$

Suppose the input space $\mathcal{X}$ is of dimension $N = 2$. Then a second-degree polynomial ($d = 2$) corresponds to the following inner product in dimension 6:

$$\forall x, x' \in \mathcal{X}, \quad K(x, x') = (x_1 x_1' + x_2 x_2' + c)^2 = \begin{bmatrix} x_1^2 \\ x_2^2 \\ \sqrt{2} x_1 x_2 \\ \sqrt{2c} x_1 \\ \sqrt{2c} x_2 \\ c \end{bmatrix} \cdot \begin{bmatrix} x_1'^2 \\ x_2'^2 \\ \sqrt{2} x_1' x_2' \\ \sqrt{2c} x_1' \\ \sqrt{2c} x_2' \\ c \end{bmatrix}.$$

The features corresponding to a second-degree polynomial are the original features (x_1, x_2), products of these features, and the constant feature.[2]

Example 6.2.6 (Gaussian kernel). Gaussian kernels are among the most frequently used kernels in applications. For any constant $\sigma > 0$, a Gaussian kernel or radial basis function (RBF) is the kernel K defined over $\mathcal{X} \subset \mathbb{R}^N$ by

$$\forall x, x' \in \mathcal{X}, \; K(x, x') = \exp\left(-\frac{\|x' - x'\|^2}{2\sigma^2}\right).$$

What mapping ϕ would lead to this kernel? Let us consider a simplification where $\sigma = 1$, let $\mathcal{X} = \mathbb{R}$, and consider the mapping $\phi(x) := \langle(\frac{1}{\sqrt{n!}} e^{-\frac{x^2}{2}} x^n)_{n \geq 0}, \cdot\rangle$. Then $K_x(\cdot) := \langle(\frac{1}{\sqrt{n!}} e^{-\frac{x^2}{2}} x^n)_{n \geq 0}, \cdot\rangle$, and

$$\langle K_x, K_{x'} \rangle = \sum_{n=0}^{\infty} \left(\frac{1}{\sqrt{n!}} e^{-\frac{x^2}{2}} x^n\right)\left(\frac{1}{\sqrt{n!}} e^{-\frac{x'^2}{2}} x'^n\right)$$

$$= e^{-\frac{x^2 + x'^2}{2}} \sum_{n=0}^{\infty} \left(\frac{(xx')^n}{n!}\right)$$

$$= e^{-\frac{(x-x')^2}{2}}.$$

Intuitively, the Gaussian kernel sets the inner product in the feature space between x, x' to be close to zero if the instances are far away from each other (in the original domain) and close to 1 if they are close.

---

2 A nice visualization using Kernel SVM with a polynomial kernel: https://www.youtube.com/watch?v= OdlNM96sHio&t=0s

Example 6.2.7 (Kernelized SVM). Recall from Theorem 5.2.8 that using the Lagrange multipliers to solve a linearly separable classification task with SVM, the solution (w_*, b_*) has the form

$$f_{w_*}(x) = \text{sign}(w_*^T x + b_*) = \text{sign}\left(\sum_{i=1}^{m} a_i y_i ((x_i)^T x) + b_* \right).$$

The "kernelized" SVM (an example can be seen in Figure 6.2) yields

$$f_{w_*}(x) = \text{sign}\left(\sum_{i=1}^{m} a_i y_i K(x_i, x) + b_* \right).$$

We simply replace the dot product $(x_i)^T x$ by $K(x_i, x)$, i. e., we compare the features through K, or, equivalently, we compare the features in some implicit higher-dimensional space where the similarity is measured through an inner product.

See Figure 6.2 for an example of Kernel SVM using a Gaussian kernel (radial basis functions, RBF).

Example 6.2.8. Consider the function $K : \mathbb{R}^2 \times \mathbb{R}^2 \to \mathbb{R}$ given as

$$K(x, y) = x_1 y_1 + x_2 y_2.$$

Is this a valid kernel, i. e., is it a symmetric, positive definite kernel?
1. Check symmetry.
2. For any $n \in \mathbb{N}$, consider the set of points $x_1, \ldots, x_n \in \mathbb{R}^2$. Verify that the matrix $\underline{K}$ given by

$$\underline{K}_{i,j} = K(x_i, x_j), \quad i, j = 1, \ldots, n,$$

is symmetric positive semidefinite.

Now that we have developed some intuition about what kernels are and how they work, let us prove the main theorem (Moore–Aronszajn theorem) of this section.

Proof of Theorem 6.2.4 (sketch). We are given a pd kernel K, and we want to construct the RKHS $\mathcal{H}$. This is done as follows.
- Let $G_1 := \{K_x : x \in \mathcal{X}\}$, where we recall that $K_x = K(x, \cdot)$.
- Let G_2 be the set of finite linear combinations of elements of G_1, i. e.,

$$G_2 := \left\{ \sum_{i=1}^{r} a_i K_{x_i} : r \in \mathbb{N}, a_i \in \mathbb{R}, x_i \in \mathcal{X}, \forall i = 1, \ldots, r \right\}.$$

We can check that G_2 is a vector space.

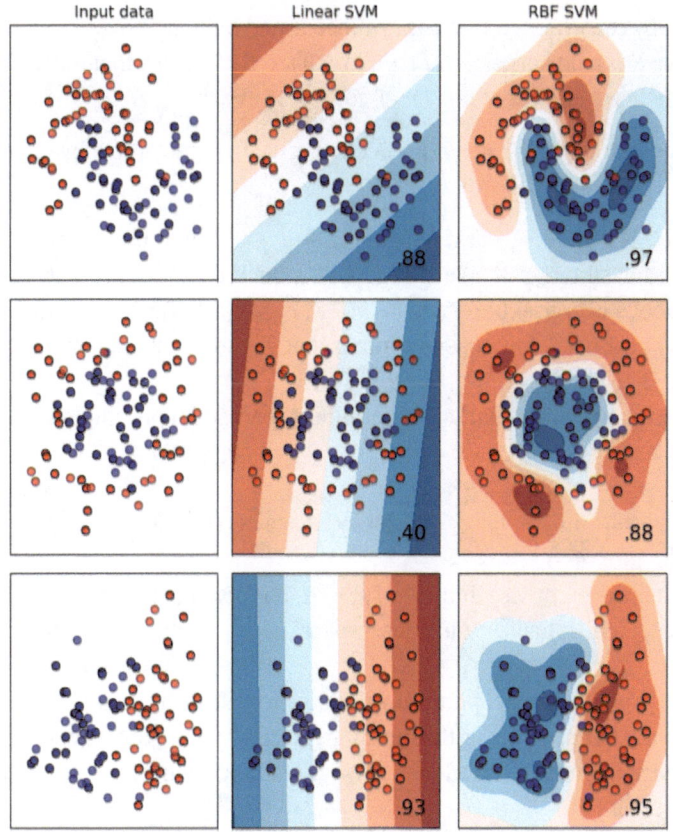

Figure 6.2: Comparison of linear SVM and SVM using an RBF kernel. Left panel: different datasets of the two-class nonseparable data are shown. Middle panel: a classifier attained using linear SVM is shown on the background; in particular, the stronger the color, the further from the classification boundary. Right panel: a classifier attained using kernel SVM (RBF kernel), showing a nonlinear separating hyperplane. The numbers in the corner show the accuracy. Source: [11].

– We define an inner product in G_2 as follows: for all $x, y \in \mathcal{X}$, define $b : G_1 \times G_1$ by $b(K_x, K_y) := K(x, y)$.
Then for all $f, g \in G_2$, there exist $r_f, r_g \in \mathbb{N}$, $\alpha_i, \beta_j \in \mathbb{R}$, and $x_i, y_j \in \mathcal{X}$ for all $i = 1, \ldots, r_f, j = 1, \ldots, r_g$ such that we can write f, g as

$$f = \sum_{i=1}^{r_f} \alpha_i K_{x_i}, \quad g = \sum_{i=1}^{r_g} \beta_i K_{y_i}.$$

We then extend b on $G_2 \times G_2$ as

$$b(f, g) = \sum_{i=1}^{r_f} \sum_{j=1}^{r_g} \alpha_i \beta_j K(x_i, y_j).$$

We readily see by construction that b is symmetric and bilinear.

Moreover, since K is a pd kernel, $b(f, f) \geq 0$ for all $f \in G_2$. To show that b is an inner product, it remains to show that $b(f, f) > 0$ for $f \neq 0$.

Exercise. Show that $|b(f, K_x)| \leq \sqrt{b(f, f) K(x, x)}$, $x \in \mathcal{X}, f \in G_2$ (Cauchy–Schwarz inequality).

Hint: look at $b(f + K_x, f + K_x)$ and $b(f - K_x, f - K_x)$ and then use the inequality $c^2 + d^2 \geq 2|cd|$ for all $c, d \in \mathbb{R}$. We choose $x \in \mathcal{X}$ such that $f(x) \neq 0$ and use the result of the above exercise: $0 < f(x)^2 = b(f, K_x)^2 \leq b(f, f) K(x, x)$. We thus have that b is strictly positive definite, which entails that b is an inner product on G_2. We then define $\langle \cdot, \cdot \rangle_{G_2} := b(\cdot, \cdot)$ and $\|f\|_{G_2} := \sqrt{\langle f, f \rangle}$.

– There is a last property that G_2 lacks to be a Hilbert space: it is not complete. We can define the space

$$\mathcal{H} := \left\{ \lim_{n \to \infty} f_n; \ (f_n)_{n \geq 1} \text{ Cauchy sequence in } (G_2, \| \cdot \|_{G_2}) \right\},$$

where the limit is the pointwise limit. For $f, g \in \mathcal{H}$, it is possible to define $\langle f, g \rangle_{\mathcal{H}} := \lim_{n \to \infty} \langle f_n, g_n \rangle_{G_2}$, which makes $\mathcal{H}$ a Hilbert space. $\qquad\square$

Remark 6.2.9. The proof of this last point turns out to be quite technical and is beyond the scope of this book. For our purpose, the intuition given by the above sketch should be sufficient. For the curious and motivated reader, we give a rough plan of how to complete the proof of the last point:

(i)   Show that if $(f_n)_{n \geq 1}$ is a $\| \cdot \|_{G_2}$-Cauchy sequence, then $f_n(x)$ is a Cauchy sequence in $\mathbb{R}$ for all $x \in \mathbb{R}$ (use the reproducing kernel and then the Cauchy–Schwarz inequality). In particular, $(f_n)_{n \geq 1}$ converges pointwise.

(ii) Show that if a Cauchy sequence $f_n \to 0$ pointwise as $n \to \infty$, then $\|f_n\|_{G_2} \to 0$ (fix $N \in \mathbb{N}$ large enough, write $\langle f_n, f_n \rangle_{G_2} = \langle f_n - f_N, f_n \rangle_{G_2} + \langle f_N, f_n \rangle_{G_2}$, and bound the two terms).

(iii) Show that for two Cauchy sequences $(f_n)_{n \geq 1}$ and $(g_n)_{n \geq 1}$ in G_2, the sequence $(\langle f_n, g_n \rangle_{G_2})_{n \geq 1}$ is a Cauchy sequence in $\mathbb{R}$. (Use the Cauchy–Schwarz inequality.)

(iv) For $f, g \in \mathcal{H}$, define $\langle f, g \rangle_{\mathcal{H}} := \lim_{n \to \infty} \langle f_n, g_n \rangle_{G_2}$ and show using (ii) that it does not depend on the choice of the Cauchy sequences $(f_n)_{n \geq 1}$ and $(g_n)_{n \geq 1}$ that converge pointwise to f and g.

(v)   Show that $\langle \cdot, \cdot \rangle_{\mathcal{H}}$ is indeed an inner product.

(vi) Show that G_2 is dense in $\mathcal{H}$.

(vii) Show that $\mathcal{H}$ is complete: take a Cauchy sequence $(f_n)_{n \geq 1}$ in $\mathcal{H}$ and use (vi) to define a sequence $(g_n)_{n \geq 1}$ in G_2 such that $\lim_{n \to \infty} \|f_n - g_n\|_{\mathcal{H}} = 0$; check that $(g_n)_{n \geq 1}$ is a Cauchy sequence in G_2 that converges pointwise to a function $g \in \mathcal{H}$ and show that f_n converges to g in $\mathcal{H}$.

(viii) Check that K is the reproducing kernel of $\mathcal{H}$.

6.3 Mercer's theorem

In this section, we state results on pd kernels without proofs. We will see that a kernel can be represented as a sum of its *eigenfunctions*, similarly to the eigendecomposition of a symmetric matrix. Thanks to this representation, the inner product in the associated RKHS can be seen as an inner product in $L^2(\mu)$, the set of square-integrable functions against some measure μ with compact support in $\mathcal{X}$. We will then use this representation to see how the inner product in the RKHS corresponds, for some specific examples, to a dot product in $\mathbb{R}^n$.

Definition 6.3.1. Let μ be a finite measure on a compact subset $B \subset \mathcal{X}$. The integral operator $T_K : L^2(\mu) \to L^2(\mu)$ induced by a pd kernel K and the measure μ is defined by

$$T_K f : \mathcal{X} \to \mathbb{R},$$

$$x \mapsto T_K f(x) := \int_{\mathcal{X}} K(x, x') f(x') \mu(dx').$$

We say that a map $e : \mathcal{X} \to \mathbb{R}$ is an *eigenfunction* of T_K with *eigenvalue* λ if $T_K e = \lambda e$.

Positive definite kernels can be seen as an infinite-dimensional generalization of positive definite matrices. We admit the following well-known fact:

A real matrix $M \in \mathbb{R}^{n \times n}$ is semipositive definite with rank $k \leq n$ if and only if $M = B^T B$ for some matrix $B^{k \times n}$ of rank k, and if M is positive definite, then $k = n$. In particular, the columns $b_1, \ldots, b_n \in \mathbb{R}^k$ of B are such that $M_{i,j} = \langle b_i, b_j \rangle$, where the inner product denotes the dot product.

It turns out that we can decompose the kernel K using the eigenfunctions of the operator T_K and get an analogue of the above fact for pd kernels. For a measure μ on a set B, let $L^2(B, \mu) := \{f : B \to \mathbb{R} \text{ measurable} : \int_B f(x)^2 \mu(dx) < \infty\}$.

Theorem 6.3.2 (Mercer's theorem). *Let K be a continuous pd kernel, and let μ be a finite measure supported on a compact subset $B \subset \mathcal{X}$. There exists an orthonormal basis $(e_i)_{i \geq 1}$ of $L^2(B, \mu)$ consisting of eigenfunctions of T_K with nonnegative eigenvalues $(\lambda_i)_{i \geq 1}$. Furthermore, for all $i \geq 1$, if $\lambda_i > 0$, then e_i is continuous, and for all $x, x' \in B$,*

$$K(x, x') = \sum_{i \geq 1} \lambda_i e_i(x) e_i(x'),$$

where the series converges uniformly on B.

Suppose that T_K has finitely many nonzero eigenvalues, say $n \in \mathbb{N}$. Then by Mercer's theorem we can write K as a dot product:

$$K(x, x') = \sum_{i=1}^{n} \lambda_i e_i(x) e_i(x') = \langle e_\lambda(x), e_\lambda(x') \rangle,$$

where $e_\lambda(x) := (\sqrt{\lambda_1} e_1(x), \dots, \sqrt{\lambda_n} e_n(x)) \in \mathbb{R}^n$. Now by Theorem 6.2.4 this means that

$$\langle \phi(x), \phi(x') \rangle_{\mathcal{H}} = \langle e_\lambda(x), e_\lambda(x') \rangle,$$

that is, the inner product in the RKHS $\mathcal{H}$ actually corresponds to a dot product in $\mathbb{R}^n$!

Let us come back to Example 6.2.3 to illustrate this.

Example 6.3.3. In Example 6.2.3, we started from a feature map ϕ_x to define the kernel $K(x,x') = x_1 x_1' + x_2 x_2' + x_1 x_2 x_1' x_2'$. Visually, we recognized the dot product in $\mathbb{R}^3$ of the vectors $(x_1, x_2, x_1 x_2)^T$ and $(x_1', x_2', x_1' x_2')^T$. We now show how to derive this from Mercer's theorem.

Let $B := [-c, c]^2 \subset \mathbb{R}^2$ with arbitrary $c > 0$. Let μ be the Lebesgue measure on B. The associated integral operator reads as

$$T_K f(x) = \int_{[-c,c]^2} (x_1 x_1' + x_2 x_2' + x_1 x_2 x_1' x_2') f(x')\, dx'.$$

We look for the eigenfunctions of T_K:

$$T_K f(x) = \lambda f(x) \quad \forall x \in B$$
$$\Leftrightarrow \quad a_1 x_1 + a_2 x_2 + a_3 x_1 x_2 = \lambda f(x), \tag{6.3}$$

where

$$a_1 := \int_{[-c,c]^2} x_1' f(x')\, dx',$$

$$a_2 := \int_{[-c,c]^2} x_2' f(x')\, dx',$$

$$a_3 := \int_{[-c,c]^2} x_1' x_2' f(x')\, dx'.$$

We can check that $e_1(x) = \frac{x_1}{\lambda_1^{1/2}}$, $e_2(x) = \frac{x_2}{\lambda_2^{1/2}}$, and $e_3(x) = \frac{x_1 x_2}{\lambda_3^{1/2}}$ are eigenfunctions with respective eigenvalues $\lambda_1 = \lambda_2 = \frac{2}{3} c^3$ and $\lambda_3 = \frac{4}{9} c^6$ Let us do it only for the first eigenfunction/eigenvector: we plug e_1 into the left-hand side of (6.3), and we get

$$\frac{1}{\lambda_1^{1/2}} \left(x_1 \int_{[-c,c]^2} (x_1')^2\, dx' + x_2 \int_{[-c,c]^2} x_1' x_2'\, dx' + x_1 x_2 \int_{[-c,c]^2} (x_1')^2 x_2'\, dx' \right)$$

$$= e_1(x) \left[\frac{(x_1')^3}{3} \right]_{-c}^{c} + 0 + 0$$

$$= \lambda_1 e_1(x),$$

as claimed.

The reader can also check that e_1, e_2, and e_3 are orthonormal, and it is clear from (6.3) that T_K has no other eigenfunction with nonzero eigenvalue (that is not a combination of these three).

Let $e_\lambda(x) := (\lambda_1^{1/2} e_1(x), \lambda_2^{1/2} e_2(x), \lambda_3^{1/2} e_3(x))^T$. Then for any $x, x' \in B$, we thus see that

$$K(x, x') = \langle \phi_x, \phi_{x'} \rangle_{\mathcal{H}} = \langle e_\lambda(x), e_\lambda(x') \rangle.$$

Conclusion

For a simple kernel, we used Mercer's theorem to express it as a dot product in $\mathbb{R}^3$ instead of an abstract inner product in the RKHS of its canonical feature map ϕ_x and $\phi_{x'}$. We thus speak of *noncanonical feature map* $\varphi(x) = (x_1, x_2, x_1 x_2)^T$ of the RKHS, for which the corresponding inner product is equivalent.

Note that we made an arbitrary choice for the compact set $B = [-c, c]^2$ and finite measure $\mu(dx) = \mathbb{1}_B(x)\, dx$. Changing c does not change the eigenfunctions (inside the smallest B), but does change the eigenvalues. Choosing a shape different from a square for B would more profoundly change the operator and then the eigenfunctions, which would lead to other noncanonical feature maps and another Hilbert space with, again, an equivalent inner product.

6.4 Representer theorem

The representer theorem plays an large role in a large class of learning problems. It provides the means to reduce an infinite-dimensional optimization problem to a tractable finite-dimensional one.

Theorem 6.4.1. *Let $\mathcal{X}$ be a set, let K be a positive definite kernel on $\mathcal{X}$, and let $\mathcal{H}$ be its corresponding RKHS. Let $S = \{(x_1, y_1), \ldots, (x_m, y_m)\}$ be a finite set of points in $\mathcal{X} \times \mathcal{Y}$. Finally, let $\lambda > 0$, and let $\ell : \mathcal{Y} \times \mathcal{Y} \to \mathbb{R}_+$ be an arbitrary loss function. Then the optimization problem*

$$\min_{h \in \mathcal{H}} \sum_{i=1}^{m} \ell(h(x_i), y_i) + \lambda \|h\|_{\mathcal{H}}^2$$

has a unique solution, which admits a representation of the form

$$\forall x \in \mathcal{X}, \quad f_w(x) = \sum_{i=1}^{m} w_i K(x_i, x) = \sum_{i=1}^{m} w_i K_{x_i}(x).$$

For a more general version of the representer theorem and references, see [38].

Proof. Let $\mathcal{H}_s$ be the subspace spanned by the canonical feature maps of the training data:

$$\mathcal{H}_s = \left\{ f \in \mathcal{H} : f(x) = \sum_{i=1}^m w_i K_{x_i}(x), (w_1, \ldots, w_m) \in \mathbb{R}^m \right\}.$$

Note that $\mathcal{H}_s$ is a finite-dimensional subspace of $\mathcal{H}$. Denoting its complement in $\mathcal{H}$ by $\mathcal{H}_s^\perp = \{ f \in \mathcal{H} : \langle f, g \rangle_\mathcal{H} = 0, \forall g \in \mathcal{H}_s \}$, for all $f \in \mathcal{H}$, there exists a unique orthogonal decomposition

$$f = \underbrace{f_s}_{\in \mathcal{H}_s} + \underbrace{f_\perp}_{\in \mathcal{H}_s^\perp}.$$

In particular, since $K_{x_i} \in \mathcal{H}_s$ for all $i = 1, \ldots, m$, we have

$$\forall i = 1, \ldots, m, \quad f_\perp(x_i) = \langle f_\perp, K_{x_i} \rangle_\mathcal{H} = 0$$

by the reproducing property of the RKHS $\mathcal{H}$. Therefore

$$\forall i = 1, \ldots, m, \quad f(x_i) = f_s(x_i),$$

and we see that the orthogonal part $f_\perp$ does not influence the loss $\ell(f(x_i), y_i)$ in the minimization problem.

We now show that if $f \in \mathcal{H}$ solves the optimization problem, then $f \in \mathcal{H}_s$. By the orthogonality of f_s and $f_\perp$ we have

$$\|f\|_\mathcal{H}^2 = \langle f_s + f_\perp, f_s + f_\perp \rangle_\mathcal{H} = \|f_s\|_\mathcal{H}^2 + \|f_\perp\|_\mathcal{H}^2.$$

As the objective function is strictly increasing in the last variable (the norm), necessarily,

$$\|f_\perp\|_\mathcal{H}^2 = 0;$$

otherwise, f would not minimize the objective function. Hence we have shown that $f \in \mathcal{H}_s$, which concludes the proof. □

6.5 Kernel (ridge) regression

In Chapter 5, we studied linear models and obtained the analytical solution of linear regression in Theorem 5.1.3 and that of linear ridge regression in Theorem 5.1.9 when considering the squared error loss. In the current chapter, we saw how kernel methods can express nonlinear relationships between inputs and outputs as linear combinations of elements of an RKHS; in particular, see Theorem 6.4.1.

Question. Can we derive an analytical solution with kernel methods?

It turns out that by considering the mean-squared loss the answer is yes. We define respectively the objective function and the regularized objective function for all $h \in \mathcal{H}$ as

$$L(h) := \frac{1}{m} \sum_{i=1}^{m} (h(x_i) - y_i)^2,$$

$$L_r(h) := L(h) + \mathcal{R}(h),$$

where $\mathcal{R}(h) := \lambda \|h\|_{\mathcal{H}}^2$ for some $\lambda > 0$. The kernel regression and kernel ridge regression problems then read as

$$\min_{h \in \mathcal{H}} L(h) \tag{6.4}$$

and

$$\min_{h \in \mathcal{H}} L_r(h). \tag{6.5}$$

Theorem 6.5.1. *Let $S = \{(x_i, y_i) \in \mathcal{X} \times \mathcal{Y}; i = 1, \ldots, m\}$ and recall that $\underline{K} = (K(x_i, x_j))_{1 \le i,j \le m}$ denotes the Gram matrix of the dataset. Let $K(x, X) = (K(x, x_1), \ldots, K(x, x_m))$.*
(i) *If $\underline{K}$ is invertible, then a solution to (6.4) is h_* given by*

$$h_*(x) = K(x, X)\underline{K}^{-1}Y.$$

Moreover, h_* is the unique solution which has the minimal RKHS norm.
(ii) *If $-m\lambda$ is not an eigenvalue of $\underline{K}$, then (6.5) is minimized at h_* given by*

$$h_*(x) = K(x, X)(\underline{K} + m\lambda I_m)^{-1}Y,$$

where I_m is the $m \times m$ identity matrix. Moreover, the solution is unique.

Proof. (i) Define

$$\mathcal{H}_s := \left\{ h(\cdot) = \sum_{i=1}^{m} w_i K(\cdot, x_i) : w_1, \ldots, w_m \in \mathbb{R} \right\}.$$

Note that

$$h_*(x) = K(x, X)\underline{K}^{-1}Y$$
$$= \sum_{i,j=1}^{m} K(x, x_i)\underline{K}_{i,j}^{-1}y_j$$
$$= \sum_{i=1}^{m} w_i K(x, x_i)$$

with $w_i := \sum_{j=1}^{m} \underline{K}_{i,j}^{-1}y_j$. This shows that $h_* \in \mathcal{H}_s$. Note also that $h_*(x_i) = y_i$ for all $i = 1, \ldots, m$, so that h_* is indeed a solution of (6.4).

Let $g \in \mathcal{H}$ be another solution of (6.4). We uniquely decompose $g = g_s + g_\perp$ with $g_s \in \mathcal{H}_s$ and $g_\perp \in \mathcal{H}_\perp$, and since $K_{x_i} \in \mathcal{H}_s$, we have for all $i = 1, \ldots, m$ that $g_\perp(x_i) = \langle g_\perp, K_{x_i} \rangle_{\mathcal{H}} = 0$. Hence we see that $g_s(x_i) = y_i$ for all $i = 1, \ldots, m$, that is, g_s is a solution of (6.4). We deduce that $\langle h_* - g_s, K_{x_i} \rangle_{\mathcal{H}} = h_*(x_i) - g_s(x_i) = 0$ for all $i = 1, \ldots, m$ since both functions are solutions, so that $h_* - g_s \in \mathcal{H}_s \cap \mathcal{H}_s^\perp$ and $h_* = g_s$. Finally, by orthogonality we have $\|g\|_{\mathcal{H}} = \|h_*\|_{\mathcal{H}} + \|g_\perp\|_{\mathcal{H}}$, which is minimal for $g_\perp = 0$ and proves that h_* is the unique solution with minimal norm $\| \cdot \|_{\mathcal{H}}$.

(ii) Thanks to the representer theorem (Theorem 6.4.1), we know that if f_w is a solution of (6.5), it can be written as

$$f_w(x) = \sum_{i=1}^{m} w_i K(x, x_i).$$

Note also that the squared RKHS norm of f_w can be written as

$$\|f_w\|_{\mathcal{H}}^2 = \langle f_w, f_w \rangle_{\mathcal{H}}$$
$$= \sum_{i,j=1}^{m} w_i w_j \langle K_{x_i}, K_{x_j} \rangle_{\mathcal{H}}$$
$$= \sum_{i,j=1}^{m} w_i w_j K(x_i, x_j),$$

thanks to the reproducing property. Hence the optimization problem (6.5) is equivalent to

$$\min_{w \in \mathbb{R}^m} \frac{1}{m} \|\underline{K}w - Y\|^2 + \lambda w^T \underline{K} w = \min_{w \in \mathbb{R}^m} \frac{1}{m} (w^T \underline{K}^2 w - 2w^T \underline{K} Y + \|Y\|^2) + \lambda w^T \underline{K} w.$$

The rest of the proof is now very similar to that of the linear ridge regression, i. e., Theorem 5.1.9. The reader can check that the problem is convex, and so there exist solutions. Taking the gradient of the function to optimize with respect to w yields

$$\frac{1}{m}(2\underline{K}(\underline{K} + m\lambda I_m)w - 2\underline{K}Y).$$

The above gradient is zero if and only if $w = (\underline{K} + m\lambda I_m)^{-1}Y$, which therefore achieves the unique global minimum. Hence the solution is given by

$$K(x, X)w = K(x, X)(\underline{K} + m\lambda I_m)^{-1}Y,$$

as claimed, which concludes the proof. $\qquad\square$

7 Gaussian processes

So far, we have considered models with a clear functional structure, meaning that we consider a class of functions with explicit relationship between inputs and outputs, for example, linear functions. Another approach to tackle the learning problem (both in regression and classification) is to give a prior probability to every possible function, where higher probabilities are given to functions that we consider to be more likely, for example, because they are smoother than other functions.

The first mentioned approach has an obvious problem in that we have to decide upon the richness of the class of functions considered; if we are using a model based on a certain class of functions (e. g., linear functions) and the target function is not well modeled by this class, then the predictions will be poor. We can increase the flexibility of the class of functions, but this runs into the danger of overfitting, where we can obtain a good fit to the training data but a bad fit on points outside of the training set.

The second approach appears to have a serious problem in that, surely, there is an uncountably infinite set of possible functions, so how are we going to compute with this set in finite time? A *Gaussian process* is a generalization of a Gaussian random variable. So far, we have seen random variables that are scalars or vectors; similarly, a stochastic process is a random function.

Example 7.0.1. Consider a one-dimensional regression problem. In Figure 7.1(a), we show a number of functions drawn at random from the prior distribution over functions specified by a particular *Gaussian process*, which favors smooth functions. This prior is taken to represent our prior beliefs over the kinds of functions we expect to observe before seeing any data.

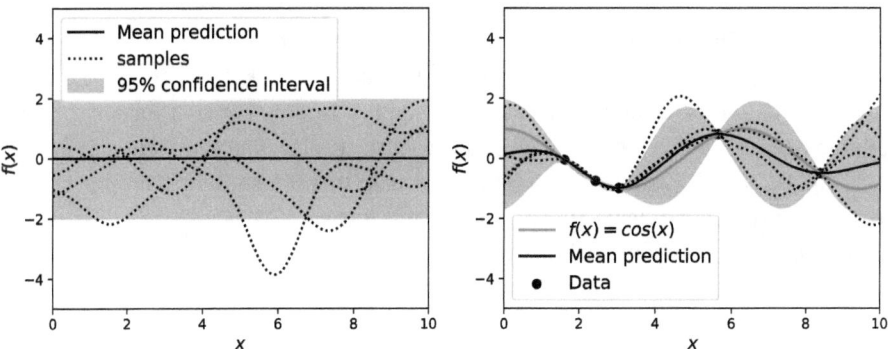

Figure 7.1: *Left*: Shows four samples drawn from a Gaussian process when no data have been observed (prior). *Right*: Shows posterior GP after five datapoints from the function $f(x) = \cos(x)$ have been observed. In both plots the mean prediction is shown as the solid line, and four samples from the posterior are shown as dotted lines. The shaded region denotes twice the standard deviation at each input value x.

https://doi.org/10.1515/9783111288994-007

Suppose that we see two datapoints (x_1, y_1) and (x_2, y_2). Then we wish to consider only functions that pass by those two points. In Figure 7.1(b), we see functions that are consistent with the observed data (dashed lines), and the solid line depicts the mean of all functions consistent with those observations. Notice how uncertainty is reduced close to the observations: this is because we have the prior that the functions are smooth.

7.1 Formal definition

7.1.1 Gaussian processes

To ease the presentation, we consider Gaussian processes with values in $\mathbb{R}$ while we allow the input space $\mathcal{X}$ to be multidimensional.

Definition 7.1.1 (Gaussian process). Let $\mathcal{X}$ be a nonempty set, let $K : \mathcal{X} \times \mathcal{X} \to \mathbb{R}$ be a positive definite kernel, and let $\mu : \mathcal{X} \to \mathbb{R}$ be a real-valued function. Then a random function $f : \mathcal{X} \to \mathbb{R}$ is called a Gaussian process (GP) with mean function μ and covariance kernel K, whose law is denoted by $GP(\mu, K)$, if for any finite set $X = (x_1, \dots, x_m) \in \mathcal{X}$ of any size $m \in \mathbb{N}$, the random vector

$$f_X = (f(x_1), \dots, f(x_m))^T \in \mathbb{R}^m$$

follows the multivariate normal distribution $\mathcal{N}(\mu_X, \underline{K}_{XX})$ with covariance matrix $\underline{K}_{XX} = (K(x_i, x_j))_{i,j=1}^m \in \mathbb{R}^{m \times m}$ and mean vector

$$\mu_X = (\mu(x_1), \dots, \mu(x_m))^T \in \mathbb{R}^m.$$

Remark 7.1.2. This definition implies that if f is a Gaussian process, then there exist a mean function $\mu : \mathcal{X} \to \mathbb{R}$ and a covariance kernel $K : \mathcal{X} \times \mathcal{X} \to \mathbb{R}$ characterizing the law of f, that is, $GP(\mu, K)$. On the other hand, it is also true that for any positive definite kernel K and mean function μ, there exists a corresponding Gaussian process $f \sim GP(\mu, K)$. Note that saying "f is a $GP(\mu, K)$" means that f is a random variable in some space of functions with law $GP(\mu, K)$; the same way, saying "X is a standard Gaussian" means that X is a random variable on $\mathbb{R}$ with mean 0 and variance 1. The one-to-one correspondence above thus states that for all (μ, K), there exists a unique law $GP(\mu, K)$ and vice versa.

Remark 7.1.3. The positive definite kernel K is equivalent to the covariance kernel/covariance function, that is, it can be written as

$$K(x, x') = \mathbb{E}_{f \sim GP(\mu, K)}[(f(x) - \mu(x))(f(x') - \mu(x'))], \quad x, x' \in \mathcal{X},$$

where the expectation is with respect to the random function $f \sim GP(\mu, K)$.

Example 7.1.4 (A concrete GP). The most common choices for $\mu(x)$ and $K(x, x')$ are $\mu(x) = 0$ and $K(x, x') = \exp(-(x - x')^2/2)$.

7.1.2 Kernel (covariance functions)

The (covariance function) kernel is a crucial ingredient in a Gaussian process predictor, as it encodes our assumptions about the function we wish to learn. From a slightly different viewpoint it is clear that in supervised learning the notion of similarity between data points is crucial; it is a basic similarity assumption that points with inputs x that are close are likely to have similar target values y, and thus training points that are near to a test point should be more informative about the prediction at that point. Under the Gaussian process view, it is the covariance function that defines nearness or similarity.

What is the effect of choosing a kernel to define the $GP(\mu, K)$? We answer this question by looking at specific examples. Note that without loss of generality, we can assume that the mean function is null since we can always recenter a Gaussian process as $f - \mu$.

Definition 7.1.5 (Squared exponential covariance function). The squared exponential (SE) covariance function has the following form for $r = x - x', x, x' \in \mathcal{X}$:

$$K_{SE}(r) = \exp\left(-\frac{r^2}{2\ell^2}\right)$$

with parameter $\ell \in \mathbb{R}^+$ defining the *characteristic length-scale*. This kernel is also commonly referred to as the *radial basis function (RBF) kernel*.

We make two observations regarding K_{SE}. The first one concerns the parameter ℓ: the covariance between two points remains close to 1 for small $r \geq 0$ and then decays quickly once r is of order ℓ, since $e^{-\frac{r^2}{2\ell^2}}$ decreases exponentially fast to 0 past that point. This explains the name "characteristic length-scale" for the parameter ℓ; see, in particular, Figure 7.2 for the graph of $K_{SE}(|r|)$ with $\ell = 2$ and Figure 7.3 for the impact of ℓ on the GP.

The second observation concerns the regularity of K_{SE}: the covariance of two nearby points with $r \approx \epsilon > 0$ small behaves as $K_{SE}(\epsilon) \approx 1 - O(\epsilon^2)$. Hence the values of $f \sim GP(0, K_{SE})$ at points that are close are highly correlated, and we can thus expect that f is likely to be a rather smooth function (we come back to this later on).

Even though the squared exponential is a widely used kernel, smoothness can be a disadvantage when representing nonsmooth processes (e. g., stock market prices). For such processes, we prefer a covariance kernel from the *Matérn class*, for which smoothness can be controlled.

Definition 7.1.6 (Matérn class of covariance functions). The Matérn class of covariance functions is given for x, x' such that $\|x - x'\| = r$ by

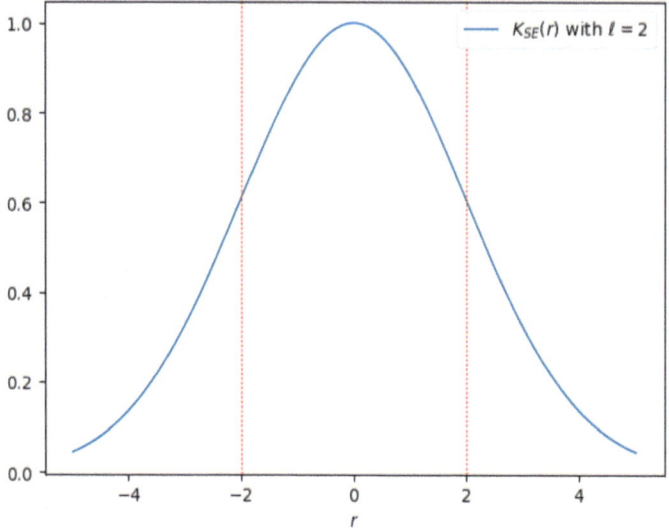

Figure 7.2: Graph of $K_{SE}(|r|)$ for $\ell = 2$.

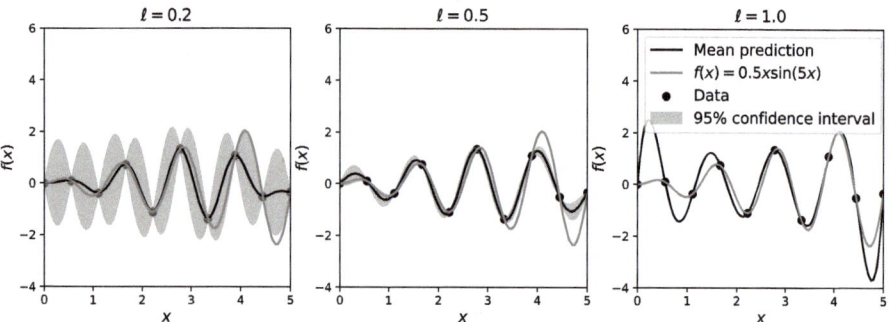

Figure 7.3: Effects of varying the hyperparameter ℓ in the squared exponential covariance function. It regulates the influence of neighboring points. If ℓ is small, then the influence of each point is more localized, shown by the uncertainty growing as soon as we are not at the observed point. As ℓ grows, the influence of each point increases.

$$K_{\text{Matern}}(r) = \frac{2^{1-\nu}}{\Gamma(\nu)} \left(\frac{\sqrt{2\nu}r}{\ell} \right)^{\nu} K_{\nu}\left(\frac{\sqrt{2\nu}r}{\ell} \right)$$

with positive parameters ν and ℓ, where K_{ν} is a modified Bessel function of the second kind [41].

The length-scale parameter ℓ plays the same role as for the squared exponential kernel, that is, it measures how far should points influence each other. The parameter ν controls the *smoothness* of the kernel; see Figure 7.4, where the map $K_{\text{Matern}}(|r|)$ is plotted for different values of ν.

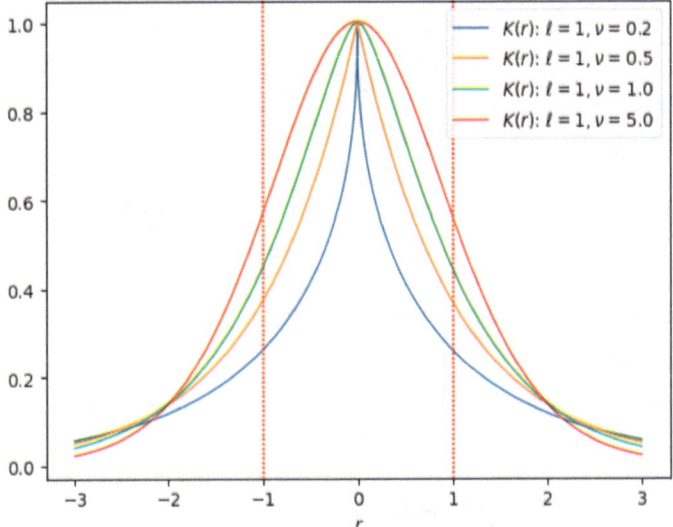

Figure 7.4: The graph of $K_{\text{Matern}}(|r|)$ is plotted for different values of v and for length-scale parameter $\ell = 1$. We see that the smaller the v, the spikier the kernel at 0.

Both covariance functions obtained from the SE and the Matérn kernels are so-called *stationary covariance functions*, as they are functions of $r = \|x - x'\|$ and thus invariant to translations in the input space. The covariance functions given above decay monotonically with r and are always positive. However, this is not a necessary condition to be a covariance function.[1] For example, the following kernel constitutes a nonstation-ary covariance function:

Definition 7.1.7 (Dot product covariance function). The dot product covariance function has the form

$$K(x, x') = \sigma_0^2 + x \cdot x', \quad x, x' \in \mathcal{X}.$$

7.2 Smoothness properties

In this section, we investigate how the covariance kernel of a Gaussian process charac-terizes its smoothness. Suppose that $(f(x))_{x \in \mathcal{X}} \sim GP(0, K_{SE})$. Then we have

$$\mathbb{E}\left[\frac{1}{\|x - x_0\|^2}(f(x) - f(x_0))^2\right] = \frac{1}{\|x - x_0\|^2}(K_{SE}(0) - 2K_{SE}(\|x - x_0\|) + K_{SE}(0)).$$

---

1 For example, a valid covariance function can have the form of a damped oscillation.

In particular, since $K_{SE}(h) = 1 - \frac{h^2}{2\ell^2} + o(h^2)$ as $h \to 0$, we see that we can take the limit as $x \to x_0$ of the expectation above, which leads to

$$\lim_{x \to x_0} \mathbb{E}\left[\frac{1}{\|x - x_0\|^2}(f(x) - f(x_0))^2\right] = \frac{1}{\ell^2}.$$

This quantity being finite, it seems to indicate that f is somewhat differentiable thanks to the differentiability of K_{SE}. This is not always true; however, this intuition can be formalized introducing the notions of *mean square continuity* and *mean square differentiability*.

To ease the presentation, we consider $\mathcal{X} = \mathbb{R}$; the multidimensional extension follows from the analogue arguments.

Definition 7.2.1. We say that a random process $(X(x))_{x \in \mathbb{R}}$ is mean square continuous at x_0 if

$$\lim_{x \to x_0} \mathbb{E}[(X(x) - X(x_0))^2] = 0.$$

We say that $(X(x))_{x \in \mathbb{R}}$ is mean square differentiable at x_0 if there exists a random process $(X'(x))_{x \in \mathbb{R}}$ such that

$$\lim_{x \to x_0} \mathbb{E}\left[\left(X'(x_0) - \frac{X(x) - X(x_0)}{x - x_0}\right)^2\right] = 0.$$

When a sequence of random variables $(Y_n)_{n \geq 1}$ is such that $\mathbb{E}[(Y_n - Y)^2] \to 0$ for some square-integrable random variable Y, we say that Y_n converges to Y in $\mathbb{L}^2$.

We see that mean square differentiability implies mean square continuity. Note that if a GP f is continuous (respectively, differentiable), then it is mean square continuous (respectively, mean square differentiable), but the converse is not true.

Example 7.2.2. To be more precise, consider the process $g : [0,1] \to \mathbb{R}$ defined by

$$g(t) := \mathbb{1}_{\{t \leq U\}},$$

where U is uniformly distributed on $(0,1)$. Then, for all $0 \leq t < t_0 \leq 1$, we see that $\mathbb{E}[(g(t) - g(t_0))^2] = \mathbb{P}[t < U < t_0] = t_0 - t$, which tends to 0 as $t \to t_0$. In particular, g is mean square continuous on $[0,1]$, but, clearly, it is not continuous.

The realization of a stochastic process is called *a path*. The example above shows that f can be mean square continuous, whereas almost surely, its path is not continuous.

For Gaussian processes, mean square continuity/differentiability are rather easy to verify. For simplicity, we focus on the stationary case $K(x, x') = K(|x - x'|)$. In particular, by symmetry the derivative at $r = 0$ is $K'(0) = 0$, so that $K''(0) = \lim_{r \to 0} K(r)/r^2$.

Theorem 7.2.3. *Let $f \sim GP(0, K)$ for some stationary kernel $K : \mathbb{R} \times \mathbb{R} \to \mathbb{R}$.*
(i) *The process f is mean square continuous if and only if K is continuous at 0.*

(ii) *The process f is mean square differentiable if and only if $K''(0)$ exists and is finite.*
(iii) *If f is mean square differentiable and $r \mapsto K(r)$ is twice differentiable on $\mathbb{R}$, then the mean square derivative f' of f has the law $GP(0, -K'')$.*

In the multidimensional case $f : \mathbb{R}^{d_{in}} \to \mathbb{R}^{d_{out}}$, statement (iii) reads as $\frac{\partial f(x)}{\partial x_i} \sim GP(0, (\frac{\partial^2 K}{\partial x_i \partial x_j})_{i,j \le d_{in}})$.

Proof. (i) By the definition of a covariance kernel we have

$$\mathbb{E}[(f(x) - f(x_0))^2] = K(0) + K(0) - 2K(|x - x_0|).$$

We see that taking the limit as $x \to x_0$, the mean square continuity of f at x_0 is equivalent to the continuity of K at 0 for all $x_0 \in \mathbb{R}$.

(ii) Let $Y_{x_0}(x) := \frac{f(x) - f(x_0)}{x - x_0}$. If f is mean square differentiable, then $Y_{x_0}(x)$ converges in $\mathbb{L}^2$ to the mean square derivative $f'(x_0)$ as $x \to x_0$, and we have

$$\mathbb{E}[Y_{x_0}(x)^2] = \frac{2K(0) - 2K(x - x_0)}{(x - x_0)^2} \xrightarrow{x \to x_0} \mathbb{E}[f'(x_0)] < \infty.$$

By symmetry we must have $K'(0) = 0$, which implies that K is twice differentiable at 0.

Suppose now that K is twice differentiable at 0. Let $(x_n)_{n \ge 1}$ be a sequence in $\mathbb{R}$ such that $x_n \to x_0$ as $n \to \infty$. The goal is to show that $(Y(x_n))_{n \ge 1}$ is a Cauchy sequence in $\mathbb{L}^2$, which entails that $Y(x_n)$ converges to some Y in $\mathbb{L}^2$.[2] We write

$$\mathbb{E}[(Y_{x_0}(x_n) - Y_{x_0}(x_{n+m}))^2] = 4\frac{K(0) - K(x_n - x_0)}{(x_n - x_0)^2} + 4\frac{K(0) - K(x_{n+m} - x_0)}{(x_{n+m} - x_0)^2}$$
$$- 2\mathbb{E}\left[\frac{(X(x_n) - X(x_0))(X(x_{n+m}) - X(x_0))}{(x_n - x_0)(x_{n+m} - x_0)}\right].$$

Assume without loss of generality that $(x_k)_{k \ge 1}$ is decreasing, so that $x_n - x_0 \ge x_{n+m} - x_0 \ge 0$. For n large enough, the first two terms yield $-2K''(0) + o(1)$ since $K'(0) = 0$ by symmetry, where $o(1)$ tends to 0 with n but is independent of m. We compute the last term of the right-hand side:

$$\mathbb{E}\left[\frac{(X(x_n) - X(x_0))(X(x_{n+m}) - X(x_0))}{(x_n - x_0)(x_{n+m} - x_0)}\right]$$
$$= \frac{K(x_n - x_{n+m}) + K(0) - K(x_{n+m} - x_0) - K(x_n - x_0)}{(x_n - x_0)(x_{n+m} - x_0)}.$$

Since $K'(0) = 0$ by symmetry of the kernel, by Taylor expansion we have

---

2 This technical claim can be seen from the fact that $\mathbb{L}^2$ is complete, which we admit without proof here.

$$K(x_n - x_{n+m}) - K(x_n - x_0)$$
$$= -K'(x_n - x_0)(x_{n+m} - x_0) + O((x_{n+m} - x_0)^2)$$
$$= -K''(0)(x_n - x_0)(x_{n+m} - x_0) + O((x_n - x_0)^2).$$

Approximating $K(0) - K(x_n - x_0)$ similarly, we get

$$\mathbb{E}\left[\frac{(X(x_n) - X(x_0))(X(x_{n+m}) - X(x_0))}{(x_n - x_0)(x_{n+m} - x_0)}\right] = -2K''(0) + o(1).$$

Again, the term in $o(1)$ tends to 0 with n but is independent of m. Hence this shows that for all $a > 0$,

$$\mathbb{E}\left[(Y_{x_0}(x_n) - Y_{x_0}(x_{n+m}))^2\right] \leq o(1),$$

that is, $(Y_{x_0}(x_n))_{n\geq 1}$ is a Cauchy sequence in $\mathbb{L}^2$ and therefore converges in $\mathbb{L}^2$ to some square-integrable random variable Y. This proves that $(f(x))_{x\in\mathcal{X}}$ is mean square differentiable for all x.

(iii) Let f' denote the mean square derivative of f. Firstly, for all $n \geq 1$ and all $x_1, \ldots, x_n \in \mathcal{X}$, we can show that the vector $(f'(x_1), \ldots, f'(x_n))$ is a centered Gaussian vector, as the limit in distribution of a centered Gaussian vector. Secondly, for all $x, y \in \mathcal{X}$ and all sequences $(x_n)_{n\geq 1}, (y_n)_{n\geq 1}$ in $\mathcal{X}$ such that $x_n \to x$ and $y_n \to y$ as $n \to \infty$, the covariance kernel κ of f' is given by

$$\kappa(x,y) = \mathbb{E}[f'(x)f'(y)] = \lim_{n,m\to\infty} \mathbb{E}\left[\frac{f(x_n) - f(x)}{x_n - x} \cdot \frac{f(y_m) - f(y)}{y_m - y}\right]$$

by the definition of mean square differentiability and since convergence in $\mathbb{L}^2$ implies convergence in $\mathbb{L}^1$. We get

$$\kappa(x,y) = \lim_{n,m\to\infty} \frac{K(x_n - y_m) + K(x - y) - K(x - y_m) - K(x_n - y)}{(x_n - x)(y_m - y)}$$
$$= \lim_{m\to\infty} \frac{K'(x - y_m) - K'(x - y)}{y_m - y}$$
$$= -K''(x - y),$$

as claimed. □

Thanks to Theorem 7.2.3, we see that $f \sim GP(0, K)$ of a stationary K is k times mean square differentiable if and only if K is $2k$ times differentiable at 0. **In particular:**
- The SE covariance function is infinitely differentiable, so that the corresponding centered GP has mean square derivatives of all orders and is therefore very smooth.
- A centered GP f with Matérn covariance kernel is k times mean square differentiable if and only if $k < \nu$. See Figure 7.5 for sample paths of a $GP(0, K_{\text{Matern}})$ with varying ν.

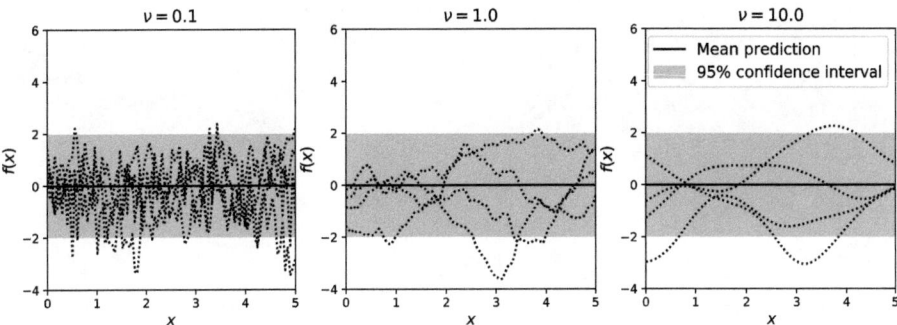

Figure 7.5: Varying the hyperparameter v regulates regularity of the functions in the GP when using the Matérn class of covariance functions. It can be shown that taking $v \to \infty$, we obtain the SE covariance function.

Let us stress once more that mean square smoothness **does not** imply smoothness of sample paths, that is, a Gaussian process f can be mean square continuous/differentiable, but $\mathbb{P}[f$ is continuous/differentiable$] = 0$ (see Example 7.2.2). Nonetheless, for stationary Gaussian processes, we have the following condition, taken from Theorem 3.4.1 of [1]. We refer the reader to this book for proofs and more details on sample path smoothness of GPs.

Theorem 7.2.4. *Let $\mathcal{X} = \mathbb{R}^d$, $d \geq 1$. Let $f \sim GP(0, K)$ be stationary for some continuous $K : \mathcal{X} \times \mathcal{X} \to \mathbb{R}$. Let $r_0 > 0$. If there exist $C > 0$ and $\epsilon > 0$ such that for all $r \in (-r_0, r_0)$,*

$$\left| K(0) - K(r) \right| \leq \frac{C}{1 + |\log(|r|)|^\epsilon},$$

then $\mathbb{P}[f$ is continuous$] = 1$.

We deduce from the above theorem that if $f \sim GP(0, K_{SE})$ on $\mathbb{R}$, then f is continuous on $\mathbb{R}$ almost surely. Indeed, we have

$$\left| K(0) - K_{SE}(r) \right| = 1 - e^{-r^2/2\ell^2} = \frac{r^2}{2\ell^2} + O(r^4),$$

which is clearly of order $o(1/|\log r|^{1+\epsilon})$ as $r \to 0$.

In Figure 7.6 the map $|\log(|r|)|^{-1.01}$ is plotted. Note that a kernel with such a spike for $K(0) - K(r)$ would be continuous almost surely by Theorem 7.2.4, which shows that the kernel needs to exhibit an extremely "spiky" behavior near zero to generate a GP that is not continuous.

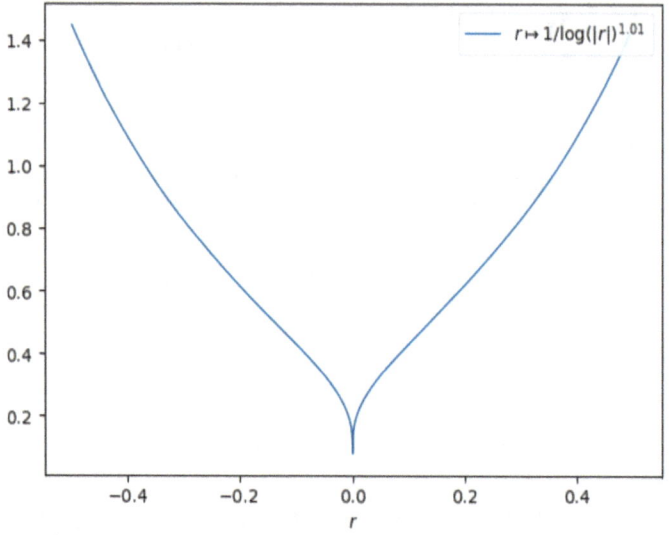

Figure 7.6: Behavior of the bound, up to constant factor, of Theorem 7.2.4 near $r = 0$ for $\epsilon = 0.01$.

7.3 Gaussian processes and kernel methods

In this short introduction to Gaussian processes, we have seen some familiar objects that were already introduced in Chapter 6 on kernel methods. In particular, for both GPs and kernel methods, the smoothness of the kernel characterizes the smoothness of the predictor. Is this a coincidence, or is there a deeper relation between GPs and kernel methods?

Recall that the Ridge regression adds a regularization penalty (scaled by λ) to the mean square error:

$$\frac{1}{m}\sum_{i=1}^{m}(y_i - f(x_i))^2 + \lambda\|f\|_{\mathcal{H}}^2.$$

By Theorem 6.5.1 the solution for the optimization above is

$$f(x) = K_{xX}(K_{XX} + m\lambda I)^{-1}Y$$

with $(K_{XX})_{i,j} = k(x_i, x_j)$, $Y = (y_1, \ldots, y_m)^T$, and $k_{xX} = (k(x, x_1), \ldots, k(x, x_m))^T$, the vector of inner products between the data and the new point x.

Bayesian posterior

To construct a predictor from a Gaussian process, we use the Bayesian approach we mentioned in Theorem 2.8.3 and the discussion below it. This approach estimates an unknown probability distribution as follows: we fix a *prior distribution* encoding our beliefs, we observe data samples, and we update our beliefs accordingly to obtain the *posterior distribution.*

Example 7.3.1. Let $\mu_* \in \mathbb{R}$ be the unknown mean of a normal distribution $D = \mathcal{N}(\mu_*, 1)$. Suppose that $S = \{x_i; i = 1, \ldots, m\}$ consists of i. i. d. samples from D. Let $P_{\text{prior}} = \mathcal{N}(0, 1)$ be our prior distribution for the mean μ, that is, our initial guess **before** observing the data S is that $\mu \sim \mathcal{N}(0, 1)$. Using S, we seek a distribution for μ that predicts better the value of μ_*. Using Bayes Theorem 2.8.3, we can formally make sense of

$$\mathbb{P}[\mu = x|S] = \frac{\mathbb{P}[S|\mu = x]\mathbb{P}[\mu = x]}{\mathbb{P}[S]}$$

$$= \frac{1}{\mathbb{P}[S]}P_{\text{prior}}[\mu = x]\prod_{i=1}^{m} e^{-\frac{(x_i-x)^2}{2}}\frac{1}{\sqrt{2\pi}}$$

$$= \frac{1}{\mathbb{P}[S]}(2\pi)^{-(m+1)/2}e^{-\frac{x^2}{2}-\sum_{i=1}^{m}\frac{(x_i-x)^2}{2}}.$$

(To be clearer, we abuse the notation and, e. g., write $\mathbb{P}[\mu = x]$ not for the actual probability that $\mu = x$, which is 0, but for the density of the law of μ at x.) The left-hand side is called the *posterior distribution*, and $\mathbb{P}[S|\mu = x] =: \ell_S(x)$ is called the *likelihood*. Hence the above is often written as

$$f_{\text{post},S}(x) \propto \ell_S(x) \times f_{\text{prior}}(x),$$

where $\propto$ denotes a proportionality relationship up to a constant in x. Let $\hat{\mu} := \frac{1}{m}\sum_{i=1}^{m} x_i$. Then the term in the exponential can be written as

$$-\frac{1}{2}\left((m+1)x^2 - 2mx\hat{\mu} + \sum_{i=1}^{m} x_i^2\right) = -\frac{1}{2/(m+1)}\left(x - \sqrt{\frac{m}{m+1}}\hat{\mu}\right)^2 + C,$$

where C does not depend on x. In particular, we have that

$$\mathbb{P}[\mu = x|S] \propto \exp\left(-\frac{(x - \frac{m}{m+1}\hat{\mu})^2}{2/(m+1)}\right).$$

This means that the posterior probability given the prior P_{prior}, and after observing the data, S is a Gaussian with mean $\sqrt{\frac{m}{m+1}}\hat{\mu}$ and variance $1/(m+1)$.

Note that the law of large numbers, that is, Theorem 2.8.1, shows that $\hat{\mu}$ converges almost surely to μ_*, so that the Bayesian posterior converges to a Dirac mass at the true value μ_*.

GP posterior

The example above shows a situation where it is possible to explicitly compute the Bayesian posterior. The general idea is that there exist true parameters θ generating observations S, and the use of the Bayes theorem can be framed as changing the viewpoint in swapping the role of θ and S, that is, considering θ as a random variable and S as parameters.

In what follows, we keep using the notation such as $\mathbb{P}[X = x]$ even for acontinuous random variable X (and conditioning on such events with probability 0). Strictly speaking, this notation is not rigorous, but it is convenient and can be rigorously justified.

The *Gaussian process regression* (also known as Kriging) is a Bayesian nonparametric method for regression. Being a Bayesian approach, it reads as follows:

- Start with a prior $GP(\mu_{\mathrm{prior}}, K_{\mathrm{prior}})$ on f_*, the unknown function generating the data.
- Observe a dataset $S = (X, Y) = \{(x_i, y_i); i = 1, \dots, m\}$ assumed to be such that $y_i = f_*(x_i) + \xi_i$, where ξ_is are i. i. d. with common law $\mathcal{N}(0, \sigma^2)$.
- Derive a posterior distribution $GP(\mu_{\mathrm{post}}, \sigma^2_{\mathrm{post}})$ for f_* based on $\mu_{\mathrm{prior}}, \sigma^2_{\mathrm{prior}}$, and S to construct the likelihood $\ell_{X,Y}(f)$, using the Bayes theorem.

Since a GP serves as a prior, the mean function μ_{prior} and the kernel K_{prior} should be chosen so that they reflect the prior knowledge or belief about the regression function f.

It is not clear what form does the Bayesian posterior

$$P_{\mathrm{post}}(f|X, Y) \propto \ell_{X,Y}(f) \times P_{\mathrm{prior}}(f)$$

take, where $P_{\mathrm{prior}} = GP(\mu_{\mathrm{prior}}, K_{\mathrm{prior}})$ and $\ell_{X,Y}(f) = \prod_{i=1}^{m} \mathbb{P}[y_i = f(x_i) + \xi_i|f]$. The next theorem provides the answer, which turns out to be a Gaussian process as well.

Theorem 7.3.2. *Assume the following:*
- *Observations are noisy, $y_i = f_*(x_i) + \xi_i$, $i = 1, \dots, m$.*
- *Noise variables are i. i. d. $\xi_i \sim \mathcal{N}(0, \sigma^2)$, $i = 1, \dots, m$.*
- *The prior is $f \sim GP(\mu, K)$.*

Denote $X = (x_1, \dots, x_m) \in \mathcal{X}^m$ and $Y = (y_1, \dots, y_m)^T \in \mathbb{R}^m$. Then the Bayesian posterior $f|(X, Y)$ has the distribution

$$f|(X, Y) \sim GP(\mu, K),$$

where $\mu : \mathcal{X} \to \mathbb{R}$ and $K : \mathcal{X} \times \mathcal{X} \to \mathbb{R}$ are given by

$$\mu(x) = \mu(x) + K_{xX}(K_{XX} + \sigma^2 I_m)^{-1}(Y - \mu_X), \quad x \in \mathcal{X},$$
$$K(x, x') = K(x, x') - K_{xX}(K_{XX} + \sigma^2 I_n)^{-1}K_{Xx'}, \quad x, x' \in \mathcal{X},$$

with $K_{Xx} = K_{xX}^T = (K(x_1, x), \dots, K(x_m, x))^T$ and $(K_{XX})_{i,j} = K(x_i, x_j)$.

Using Theorem 7.3.2, the following equivalence holds for GP-regression and kernel Ridge regression: We have $\mu = f_{KRR}$ if $\sigma^2 = m\lambda$, where

1. μ is the posterior mean function of GP-regression based on (X, Y), the GP prior $f \sim GP(0, K)$, and the modeling assumption where the noise is i. i. d. $\mathcal{N}(0, \sigma^2)$.
2. f_{KRR} is the solution to kernel ridge regression based on (X, Y), the RKHS $\mathcal{H}$, and the regularization constant $\lambda > 0$.

We refer to [30] (see Theorem 3.1 therein) for the proof of Theorem 7.3.2 and, more generally, for an extensive treatment on connections between Gaussian processes and kernel methods. It is worth noting that their proof does not rely on computing the Bayesian posterior.

In the noise-free case, we also have the correspondence with the kernel predictor:

Theorem 7.3.3. *Using the same notation as in Theorem 7.3.2, assume that*
- *Observations are noiseless,* $y_i = f_*(x_i),\ i = 1, \ldots, m$.
- *The prior is* $f \sim GP(\mu, K)$.
- *The matrix* K_{XX} *is invertible.*

Then the Bayesian posterior $f|(X, Y)$ *has the distribution*

$$f|(X, Y) \sim GP(\mu, K),$$

where $\mu : \mathcal{X} \to \mathbb{R}$ *and* $K : \mathcal{X} \times \mathcal{X} \to \mathbb{R}$ *are given by*

$$\mu(x) = \mu(x) + K_{xX}K_{XX}^{-1}(Y - \mu_X), \quad x \in \mathcal{X},$$
$$K(x, x') = K(x, x') - K_{xX}K_{XX}^{-1}K_{Xx'}, \quad x, x' \in \mathcal{X}.$$

Remark 7.3.4. The explicit solution is available for Gaussian processes; however, one of the disadvantages is the fact that the covariance matrix size scales with relation m^2 for a dataset of size m, and furthermore, inverting the covariance matrix takes approximately $O(m^3)$ operations.

7.4 Implementation details

In this section, we provide code snippets used to generate the figures of this chapter. In Listing 7.1, we show the code used to generate Figure 7.1.

Listing 7.1: Regression using a Gaussian process

```
1 from sklearn.gaussian_process import
      GaussianProcessRegressor
2 from sklearn.gaussian_process.kernels import RBF
3 import numpy as np
4 from matplotlib import pyplot as plt
5
```

```
 6 # Generate data
 7 X = np.linspace(0,10,100).reshape(-1, 1)
 8 y = np.squeeze(np.cos(X))
 9 n_samples = 4
10
11 kernel = RBF(length_scale=1.0, length_scale_bounds=(1e-2,
       1e2))
12 gaussian_process = GaussianProcessRegressor(kernel=kernel)
13
14 # Untrained Gaussian process
15 prior_mean_prediction, prior_std_prediction =
       gaussian_process.predict(X, return_std=True)
16 prior_y = gaussian_process.sample_y(X,n_samples) #
       Sampling from prior Gaussian Process at X
17
18 # Generating training data
19 rng = np.random.RandomState(1)
20 training_indices = rng.choice(np.arange(y.size), size=5,
       replace=False)
21 X_train, y_train = X[training_indices], y[training_indices
       ]
22
23 # Fit training data
24 gaussian_process.fit(X_train, y_train)
25 posterior_mean_prediction, posterior_std_prediction =
       gaussian_process.predict(X, return_std=True)
26 posterior_y = gaussian_process.sample_y(X,n_samples)
```

For the plotting routines, we refer to the code repository https://github.com/hanveiga/tmml.

8 Deep learning

In the recent years, machine learning has become often identified as *Neural Networks* or *Deep Learning*, so we could not leave this class of models out of our exposition about machine learning.

In this chapter, we start with the simplest type of deep neural networks, the fully connected (or dense) neural networks (Section 8.1). We will see that some notions of learning theory that we have encountered will be challenged, and new mathematical theory is necessary to understand why neural networks seem to work so well in practice.

Furthermore, neural networks are (by far) the models presented in these lecture notes which require more care when being set-up and trained. We will see that there are more hyperparameters, engineering choices, and ways to successfully/unsuccessfully train a neural network. We consolidate some practical advice, *tricks of the trade*, in Section 8.6.

8.1 Fully connected dense neural networks

Definition 8.1.1. Let us introduce the function (a unit)

$$a(x) = \sigma(w^T x + b),$$

where $x \in \mathbb{R}^n$ are inputs, $w \in \mathbb{R}^n$ are weights, $b \in \mathbb{R}$ is the bias, and $\sigma : \mathbb{R} \to \mathbb{R}$ is an activation function applied componentwise.

We can write different models we have studied by considering different activation functions:

- Linear regression: $\sigma(z) = z$.
- Binary classification: $\sigma(z) = \text{sign}(z)$.
- Logistic regression: $\sigma(z) = \frac{1}{1+e^{-z}}$.

A neural network is a *combination* of (typically, a lot of) these units. For example, a 1-layer neural network $f : \mathbb{R}^d \to \mathbb{R}^{d_1}$ is given by

$$f(x) = \sigma_1(W_1 x + b_1)$$

for $x \in \mathbb{R}^d$, a matrix $W_1 \in \mathbb{R}^{d_1 \times d}$, and $b_1 \in \mathbb{R}^{d_1}$.

What does $f(x)$ do? It takes a vector x and applies a linear transformation to it (by W_1, b), which returns a vector of dimension d_1, and σ is applied componentwise to this vector.

A 2-layer neural network $f : \mathbb{R}^d \to \mathbb{R}^{d_2}$ is given as

$$f(x) = \sigma_2(W_2 \sigma_1(W_1 x + b_1) + b_2),$$

where a matrix $W_2 \in \mathbb{R}^{d_2 \times d_1}$, and $b_2 \in \mathbb{R}^{d_2}$.

https://doi.org/10.1515/9783111288994-008

This leads to the recursive definition of an L-layer neural network $f : \mathbb{R}^d \to \mathbb{R}^{d_L}$:

Definition 8.1.2. A neural network is a *fully connected feedforward neural network* if its **architecture** has the following hyperparameters:
- The number of layers $L \in \mathbb{N}$;
- The activation functions $\sigma_i : \mathbb{R} \to \mathbb{R}$ $(i = 1, \ldots, N)$;
- The numbers of neurons in the lth layer $d_0, \ldots, d_L \in \mathbb{N}$, $\ell = 0, \ldots, L$;

and its output is given by

$$f(x) = \sigma_L(W_L\sigma_{L-1}(\cdots(W_2\sigma_1(W_1x + b_1) + b_2)\cdots) + b_L),$$

where $W_i \in \mathbb{R}^{d_{\ell+1} \times d_\ell}$ and the σ_is are applied componentwise, that is,

$$\sigma_i\begin{pmatrix} z_1 \\ \vdots \\ z_{d_i} \end{pmatrix} = \begin{pmatrix} \sigma_i(z_1) \\ \vdots \\ \sigma_i(z_{d_i}) \end{pmatrix} \quad \text{for } z \in \mathbb{R}^{d_i}.$$

Alternatively, a fully connected feedforward neural network can be defined as follows: let $a^{(0)}(x) := x$ for all $x \in \mathbb{R}^{d_0}$, and for $\ell = 1, \ldots, L$, define the recursion

$$\begin{aligned} \tilde{a}^{(\ell)}(x) &= W_\ell a^{(\ell-1)}(x) + b_\ell, \\ a^{(\ell)}(x) &= \sigma_\ell(\tilde{a}^{(\ell)}(x)). \end{aligned} \tag{8.1}$$

The neural network output for an input x is obtained as $a^{(L)}(x)$. The map $\tilde{a}^{(\ell)}$ is called the *preactivation* at layer ℓ, and $a^{(\ell)}$ is called the *postactivation* at layer ℓ.

Not the following properties of this definition:
- We can do both classification and regression, depending on the activation function.
- Feature selection is "built-in": if σ_L is the identity, then the neural network output is $W_L a^{(L-1)} + b_L$, which is a linear map of the penultimate layer output.
- The number of degrees of freedom (i. e., trainable parameters) of a neural network is given by

$$\text{dof} = \sum_{l=1}^{L} d_\ell d_{\ell-1} + d_\ell.$$

Remark 8.1.3. There are some engineering choices: L, σ_i, d_i. This fixes the number of degrees of freedom we have to find for W_i and b_i.

Example 8.1.4. What if all σ are the identity functions?

$$W_L(\cdots(W_1x + b_1)\cdots) + b_L = (W_L \cdots W_1)x + (b_L + W_L b_{L-1} + \cdots + W_L \cdots W_2 b_1).$$

This is just a linear predictor.

8.1.1 Loss functions

Similarly to other supervised learning models we have seen, neural networks are trained in the ERM/Regularized ERM framework.

When considering *regression-type problems*, a typical loss function to be considered is the mean squared error (or mean absolute error). For a neural network f_w, we get

$$C(w) = L(f_w) = \frac{1}{2m} \sum_{i=1}^{m} \|f_w(x_i) - y_i\|_2^2.$$

From the definition of fully connected neural networks we see that (as soon as there is a nonlinear activation) the map $w \mapsto f_w$ is nonlinear. As a consequence, even when the chosen loss L on the function space is convex, the cost C on the parameter space needs not be. In particular, for neural networks, there is no guarantee that local minima are global minima, which makes the success of training through gradient descent not obvious.

For a **classification task**, like previously, we cannot use gradient descent on nondifferentiable functions. So instead of the 0–1 loss $\ell(y, y') = \mathbb{1}_{\{y \neq y'\}}$, we consider a surrogate loss, similar to what we did with the perceptron algorithm in Section 5.2.

Consider a fully connected neural network $f_w(x) = W_L a^{(L-1)}(x) + b_L$. To perform binary classification, we can choose the predictor $\mathrm{sign}(f_w(x))$. A commonly used surrogate loss is the logistic loss given by

$$\ell(f_w(x), y) = \log(1 + e^{-y f_w(x)}).$$

This loss makes sense, since if $y = 1$ and $f_w(x) > 0$ (or $y = -1$ and $f_w(x) < 0$), then the value $\exp(-y f_w(x))$ is small, and we have $\ell(f_w(x), y) \approx 0$. Otherwise, in case of misclassification, we have $\log(1 + e^{-f_w(x)y}) > \log 2$.

Another common way to do binary classification is to choose $d_{L-1} = 1$ and $\sigma_L(z) = \mathrm{sigmoid}(z) = \frac{1}{1+\exp(-z)}$, which returns a number in $[0, 1]$, that can be interpreted as a probability that the label is 1 according to our predictor. Consider the two classes 0 and 1, then we can predict the label 0 if $f_w(x) \leq 0.5$ and 1 otherwise. The *cross-entropy loss* is then defined by

$$L(f_w) = \frac{1}{m} \sum_{i=1}^{m} (y_i \log f_w(x) + (1 - y_i) \log(1 - f_w(x))).$$

Note that if $y = 0$, then the closer $f_w(x)_{y_i}$ is to 0, the smaller the loss, and similarly if $y = 1$, then we want $f_w(x)$ close to 1. Using the convention $0 \log 0 = 0$, we see that a perfect classifier achieves the zero loss.

For multiclass classification with n classes, the above is extended by considering $d_{L-1} = n$ and $\sigma_L(z) = \mathrm{softmax}$, which returns a probability on a finite set as follows:

$$f_w(x) = \frac{1}{\sum_{i=1}^{n} \exp(a^{(L-1)}(x)_i)} \begin{pmatrix} \exp(a^{(L-1)}(x)_1) \\ \vdots \\ \exp(a^{(L-1)}(x)_n) \end{pmatrix}.$$

Then the loss function is the categorical cross-entropy

$$L(f_w) = \frac{1}{m} \sum_{i=1}^{m} \sum_{j=1}^{n} (y_{i,j} \log f_w(x_i)_j + (1 - y_{i,j}) \log(1 - f_w(x_i)_j)),$$

where we write $y_i = (y_{i,1}, \ldots, y_{i,n})$ for the vector giving the class of the ith datapoint. Since a datapoint x_i only belongs to one class, the vector y_i is made of zeros everywhere, except for a one at the component corresponding to the class x_i belongs to. Again, a perfect classifier achieves the zero loss.

8.2 Back propagation

8.2.1 Definition

Using a gradient-based optimization method, we must find a way to update each of the $\sum_{i=1}^{L} d_\ell(d_{\ell-1} + 1)$ parameters of the model. We want an efficient way to compute all these derivatives. The back-propagation algorithm can be thought of as a table-filling algorithm that takes advantage of storing intermediate results.

Example 8.2.1. Let us consider a 1-hidden layer neural network with a two-dimensional input space $f_w : \mathbb{R}^2 \to \mathbb{R}$, written explicitly as

$$f_w(x) = \sigma_1(W_1 \sigma_0(W_0 x + b_0) + b_1),$$

where $W_0 = \begin{bmatrix} w_{11} & w_{12} \\ w_{21} & w_{22} \end{bmatrix} \in \mathbb{R}^{2 \times 2}$, $W_1 = \begin{bmatrix} w_{11}^1 & w_{12}^1 \end{bmatrix} \in \mathbb{R}^{1 \times 2}$, $b_0 = \begin{bmatrix} b_1 \\ b_2 \end{bmatrix} \in \mathbb{R}^2$, and $b_1 = b_1^1 \in \mathbb{R}$.
The cost function, considering a squared loss, reads as

$$C(w) = L(f_w) = \sum_{i=1}^{m} (y_i - f_w(x_i))^2$$

$$= \sum_{i=1}^{m} (y_i - \sigma_1(W_1 \sigma_0(W_0 x + b_0) + b_1))^2.$$

To find the weights w_{ij}^1 in the weight matrix W_1, for example, we want to compute $\partial_{w_{ij}^1} C(w)$. Using the chain rule, this is given by

$$\partial_{w_{ij}^1} C(w) = \frac{\partial z_i^1}{\partial w_{ij}^1} \sigma_1'(z_i^1) \frac{\partial C(w)}{\partial \sigma_1}$$

with $z^1 = W_1 \sigma_0(x) + b_1$, so $\frac{\partial z_i^1}{\partial w_{ij}^1} = \sigma_{0,j}$. For the weight w_{ij} (belonging to W_0), we have

$$\partial_{w_{ij}} C(w) = \frac{\partial z_i}{\partial w_{ij}} \sigma_0'(z_i) \frac{\partial z_i^1}{\partial \sigma_0} \sigma_1'(z_i^1) \frac{\partial C(w)}{\partial \sigma_1}$$

with $z = W_0 x + b_0$. Note that some terms of this gradient have already been computed when we wrote the update for w_{ij}^1.

For simplicity, let us denote by W the extended weight matrix given by $W = [W; b]$, where the extra column is the bias, and the feature vector is given by $X = [X, 1]$ with the last entry given by 1. We can write a neural network recursively as

$$Z_n = W_n X_{n-1},$$
$$X_n = \sigma_n(Z_n),$$

where Z denotes the vector of z_i, and W is the (extended) matrix of weights w_{ij}. Furthermore, X_0 denotes the input vector.

Suppose we have computed $\frac{\partial C}{\partial X_{n,i}}$ for $i = 1, \ldots, d_n$ (the width of the layer n). Then we can compute recursively the following gradients:

$$\frac{\partial C}{\partial z_{n,i}} = \frac{\partial \sigma_n(z)}{\partial z}\bigg|_{z=z_{n,i}} \frac{\partial C}{\partial X_{n,i}},$$

$$\frac{\partial C}{\partial w_{ij}^n} = X_{n-1,j} \frac{\partial C}{\partial z_{n,i}},$$

$$\frac{\partial C}{\partial X_{n-1,j}} = \sum_i w_{ij}^n \frac{\partial C}{\partial z_{n,i}}.$$

This can also be written in the matrix–vector notation:

$$\frac{\partial C}{\partial Z_n} = \sigma_n'(Z_n) \circ \frac{\partial C}{\partial X_n},$$

$$\frac{\partial C}{\partial W_n} = \left(\frac{\partial C}{\partial Z_n} \right)(X_{n-1})^T,$$

$$\frac{\partial C}{\partial X_{n-1}} = W_n^T \frac{\partial C}{\partial Z_n}.$$

Note that $\sigma_n'(Z_n) = \partial_z \sigma_n(Z_n)$ produces a vector of derivatives of the activation function with respect to z, evaluated at Z_n, and $\circ$ gives the Hadarmard product between two vectors.[1]

---

1 If $a = (a_1, \ldots, a_n)$ and $b = (b_1, \ldots, b_n)$, then $a \circ b = (a_1 b_1, \ldots, a_n b_n)$.

To make things more clear for the next exposition, we can write

$$\frac{\partial C}{\partial W_n} = \left(\frac{\partial C}{\partial Z_n}\right)(X_{n-1})^T$$

$$= \left(\sigma'_n(Z_n) \circ \frac{\partial C}{\partial X_n}\right)(X_{n-1})^T$$

$$= \left(\sigma'_n(Z_n) \circ W_{n+1}^T \frac{\partial C}{\partial Z_{n+1}}\right)(X_{n-1})^T$$

$$= \left(\sigma'_n(Z_n) \circ W_{n+1}^T \sigma'_{n+1}(Z_{n+1}) \circ W_{n+2}^T \ldots \sigma'_L(Z_L) \circ \frac{\partial C}{\partial X_L}\right)(X_{n-1})^T.$$

We can write the term inside the parenthesis as

$$\delta^n := \sigma'_n(Z_n) \circ W_{n+1}^T \sigma'_{n+1}(Z_{n+1}) \circ W_{n+2}^T \ldots \sigma'_L(Z_L) \circ \frac{\partial C}{\partial X_L}.$$

Then the gradient reads

$$\frac{\partial C}{\partial W_n} = \delta^n X_{n-1}^T. \tag{8.2}$$

We can compute δ^{n-1} recursively by

$$\delta^L = \sigma'_L(Z_L) \circ \frac{\partial C}{\partial X_L}, \tag{8.3}$$

$$\delta^{n-1} = \sigma'_{n-1}(Z_{n-1}) \circ W_{n-1}^T \delta^n, \quad n = 1, \ldots, L. \tag{8.4}$$

8.2.2 Exploding and vanishing gradients

Neural networks are usually trained with gradient-based algorithms, the gradient being computed with backpropagation. Since they are compositions of functions, this may cause gradients to explode or vanish at early layers. Let us first look at the chain rule (for simplicity, in the scalar case $w \in \mathbb{R}$) for the derivative of a map $f_L(w)$ defined recursively using maps g_ℓ, $\ell = 1, \ldots, L$, by $f_0(w) = w$ and $f_{\ell+1}(w) = g_{\ell+1}(f_\ell(w))$ for all $\ell = 0, \ldots, L - 1$. We have

$$f'_L(w) = g'_L(f_{L-1}(w))f'_{L-1}(w)$$

$$= \cdots$$

$$= \prod_{i=0}^{L-1} g'_{L-i}(f_{L-i-1}(w)).$$

We see that the derivative of the composition of L maps is a product of L derivatives. In particular, if each of them is of order, say, $a \in (0, \infty)$, then $f'_L(w)$ is of order a^L. For large L, if $a < 1$, then we get a very small gradient, whereas if $a > 1$, then it can grow very large. Gradients that are too small or too large hinder training, for similar reasons to that of the learning rate we discussed in Chapter 3; see Figure 3.5. Loosely speaking, with small gradients, training gets stuck and takes too long to converge, and with large gradients, training jumps over minima and gets away from good solutions.

Recall that the backpropagation update at layer $n \in \{1, \ldots, L\}$ in a fully connected neural network is given in (8.2) in terms of δ^n, which is recursively defined from the last layer to the previous ones in (8.3). In particular, the further we go backward, the more terms in the product, which, as we said, can cause training instability.

Note that since the derivatives of the activation functions σ_ℓ appear in product (8.3), the phenomena of exploding and vanishing gradients are linked to the choices of these functions. For example, the sigmoid function $\sigma : x \mapsto (1 + e^{-x})^{-1}$ has a vanishing derivative away from some interval centered at 0. On the other hand, the derivative of a polynomial activation function of degree at least 2 explodes far enough from 0.

8.2.3 Common initialization schemes

In view of the previous section, we may wonder how to successfully train neural networks. In Section 8.6.2, we present implementation tricks based on heuristics to stabilize training. Here we are rather interested in the influence of initialization on the gradient.

Indeed, the way the neural network weights are initialized prior to training has a crucial effect on the success of training. Suppose that weights are all initialized to zero. Then the updates for the weights are given by

$$\frac{\partial C}{\partial W_n} = \delta^n X_{n-1}^T = 0,$$

as δ^n yields a zero vector, meaning that the weights will not change during training. More generally, if the weights are all initialized to a constant value c, then the update yields

$$\frac{\partial C}{\partial W_n} = \delta^n X_{n-1}^T = \delta^n \sigma_{n-1} (c\mathbf{1}^T X_{n-2}\mathbf{1})^T = \vec{\beta}\vec{y}^T$$

$$= c_n \mathbf{1}_{d_n \times d_{n-1}},$$

where $\vec{\beta}$ and $\vec{y}$ are two constant vectors, and $\mathbf{1}$ denotes the vector of 1s, and $\mathbf{1}_{d_n \times d_{n-1}}$ is the matrix of 1s. In particular, all weights at the same layer n receive the same update $c_n \in \mathbb{R}$.

In general, it is best to initialize the weights at random (independent) values, which in particular avoids this problem. Commonly used initialization schemes were born from heuristics to fix the issue of exploding and vanishing gradients we discussed in the previous section. They are based on the heuristic approximation

$$\text{Var}[X_n] \approx \sigma_n'(0)\,\text{Var}[W_n]^T\,\text{Var}[X_{n-1}],$$

which can be obtained by using a Taylor expansion of σ_n at 0, up to assuming no bias, independence of all weights, centered inputs, and $\sigma(0) = 0$. In particular, choosing $\text{Var}[w_{ij}^n] = \frac{1}{\sigma'(0)^2 d_{n-1}}$ for individual weights implies that the variance of X_n is constant across layers n.

Note that the above heuristic is valid only for the first step of gradient descent, after which weights are no longer independent. For a more rigorous approach to the initialization scale of weights, see Section 8.5 and in particular Section 8.5.4.

What does this tell us?

The above argument implies that the variance of the weights should scale as the inverse of the width times a factor that depends on the activation σ. This lead to the following famous initialization schemes:

(init i) Xavier: $w_{ij}^n \sim \text{Unif}(0, 1/\sqrt{d_{n-1}})$. This is heuristically justified as above for the tanh activation function.

(init ii) Le Cun: $w_{ij}^n \sim \mathcal{N}(0, 1/d_{n-1})$.

(init iii) He: $w_{ij}^n \sim \mathcal{N}(0, 2/d_{n-1})$. It can be motivated similarly as above for the ReLU activation.

Biases are initialized to 0, and all weights are mutually independent. Other initialization schemes, such as Glorot initialization, scale the variance by $d_{n-1} + d_n$.

8.3 Approximation theorems

In this section, we give a brief introduction to some approximation results using neural networks. This is a very active research area, but we will focus on one fundamental result, frequently cited in talks and literature as one of the earliest of its kind. Guilhoto [25] has a good introduction, focusing also on the necessary tools from analysis to understand the proofs.

The following theorem by Cybenko [12] is often called the universal approximation theorem for neural networks. We state a weaker version of the theorem proved in that paper.[2]

Theorem 8.3.1. *Let σ be the sigmoid function, that is, $\sigma(x) = \frac{1}{1+e^{-x}}$. Then finite sums of the form*

---

2 In the original paper, the theorem needs only σ to be a discriminatory function.

$$G(x) = \sum_{j=1}^{N} a_j \sigma(w_j^T x + b_j)$$

are dense with respect to the supremum norm in the space $C(I_n)$ of continuous functions on I_n, the n-dimensional unit cube $[0,1]^n$. In other words, for any $f \in C(I_n)$ and $\epsilon > 0$, there is a sum $G(x)$ of the above form for which

$$|G(x) - f(x)| < \epsilon \quad \forall x \in I_n.$$

Proof. We admit without proof the following property (called the discriminatory property) of the sigmoid: Let μ be a finite regular signed Borel measure. If

$$\int_{I_n} \sigma(w_j^T x + \theta) d\mu(x) = 0 \quad \forall (\vec{y}, \theta) \in \mathbb{R}^d \times \mathbb{R},$$

then $\mu = 0$.

Let

$$S = \left\{ f(x) = \sum_{j}^{N} a_j \sigma(w_j^T x + b_j) : N \in \mathbb{N}, a_j, w_j, b_j \in \mathbb{R}, j = 1, \ldots, N \right\}.$$

This is a linear subspace of $C(I_h)$. If the closure of S is $\bar{S} = C(I_n)$, then we are done.

The proof uses two well-known theorems of functional analysis, the Hahn–Banach theorem and the Riesz representation theorem. Assume that $\bar{S} \neq C(I_n)$. By the Hahn–Banach theorem there exists a bounded linear form $L \neq 0$ on $C(I_n)$ such that $L(\bar{S}) = 0$. By the Riesz representation theorem there exists a signed regular nonzero (as $L \neq 0$) Borel measure μ such that

$$L(h) = \int_{I_n} h(x) d\mu(x) \quad \forall h \in C(I_n).$$

Taking $h \in S$, we have $L(h) = 0$, which implies $\mu = 0$ because of the discriminatory property, which yields a contradiction. $\square$

The above theorem tells us that neural networks can approximate an arbitrary continuous function, similarly to the Stone–Weierstrass theorem for polynomials. It is worth noting that this fact is true for deeper networks as well: conditioning on the output of layer $L - 2$ and considering it as the input layer, the layers $L - 2$, $L - 1$, and L can be seen as a two-layer neural network, and we can directly apply the above theorem.

There is also a similar result by Hornik [28]. The result above also extends to classification tasks. The same result also holds for more general sigmoidal functions and the ReLU function.

8.4 Beyond feed forward neural networks

8.4.1 Convolutional neural networks

Convolutional neural networks (CNNs) are designed in such a way that they can take into account the *spatial* structure of the input. They were inspired by mice visual system and were originally designed to work with images. Compared to fully connected neural networks, CNNs have much fewer parameters since there are less connections between layers, which makes it possible to efficiently train very deep architectures.

Typically, the inputs of a convolutional neural network are *tensors*. Tensors are a generalization of matrices: a matrix is an element in $\mathbb{R}^{n_{row} \times n_{column}}$, whereas a tensor belongs to $\mathbb{R}^{n_1 \times \cdots \times n_p}$ for an arbitrary $p \in \mathbb{N}$. In the case of convolutional neural network, the input space is characterized by two dimensions $d, n \in \mathbb{N}$ and a *depth* $k \in \mathbb{N}$. For example, if the inputs are speeches, then $d = 1$, for photos, $d = 2$, and for medical imaging, $d = 3$, etc. Consider images, that is, $d = 2$. The integer n gives the size of the inputs, so that $n \times n$ is the number of pixels in an image (or $n \times m$ with $m \in \mathbb{N}$ for rectangular pictures; in this section, we assume that we have size n in each dimension d for simplicity.). The color of each pixel is characterized by the three values RGB (red, green, blue), so that the depth $k = 3$. The depth is also commonly called the number of **channels**. Given $d, n,$ k, the input space corresponds to $\mathcal{X} = (\mathbb{R}^{n^d})^k$.

A convolutional neural network typically has the structure depicted in Figure 8.1.

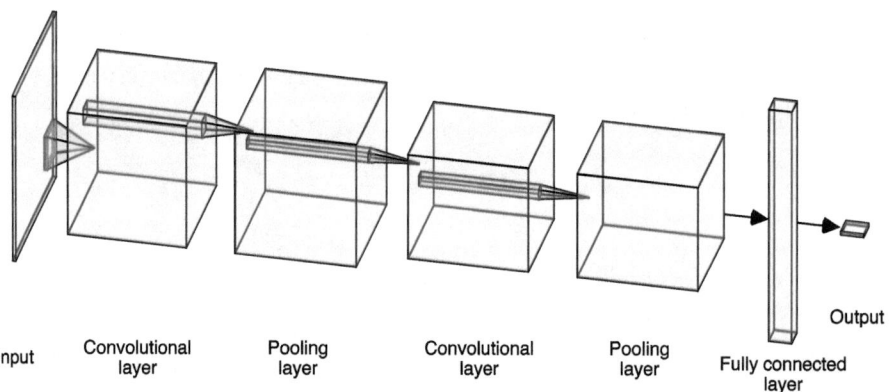

Input   Convolutional layer   Pooling layer   Convolutional layer   Pooling layer   Fully connected layer   Output

Figure 8.1: Diagram of CNN.

Below, even though the inputs are not necessarily images, we call each element at location $\mathbf{i} = (i_1, \ldots, i_d) \in \{1, \ldots, n\}^d$ a pixel. Typical layers of a convolutional network are the following:

- Convolution layer: Let $n_{in}, d_{in}, k_{in} \in \mathbb{N}$ be the size, dimension, and depth of the inputs of the convolutional layer. A *filter* (also commonly called a *kernel*) is an element of $(\mathbb{R}^{(n_{\mathrm{fil}})^{d_{\mathrm{in}}}})^{k_{\mathrm{in}}}$ that is convolved with the inputs. The size of the filter n_{fil} is smaller than the size of the inputs n_{in}; often, it is chosen equal to 3, 4 or 5. The filter $g \in (\mathbb{R}^{(n_{\mathrm{fil}})^{d_{\mathrm{in}}}})^{k_{\mathrm{in}}}$ is applied to an input $x \in (\mathbb{R}^{n_{\mathrm{in}}^{d_{\mathrm{in}}}})^{k_{\mathrm{in}}}$ as follows: denoting by g_j the jth depth (channel) of the tensor, for each depth j, we perform a discrete convolution between the filter g_j and the input x_j by sliding g_j over x_j (see Figure 8.2), followed by a sum over the channels. After the filter g is applied to x, we obtain $y \in \mathbb{R}^{(n_{\mathrm{in}}-n_{\mathrm{fil}}+1)^{d_{\mathrm{in}}}}$ such that for all $i \in \{1, \ldots, n_{\mathrm{in}} - n_{\mathrm{fil}} + 1\}^{d_{\mathrm{in}}}$, the value of y at location i is given by

$$y(\mathbf{i}) = (x * g)(\mathbf{i}) = \sum_{j \le k_{\mathrm{in}}} \sum_{p_1,\ldots,p_{d_{\mathrm{in}}}=1}^{n_{\mathrm{fil}}} x_j(\mathbf{i} + \mathbf{p} - \mathbf{1})g(\mathbf{p}),$$

where $\mathbf{p} = (p_1, \ldots, p_{n_{\mathrm{fil}}})$ is the position of the filter pixel, and $\mathbf{1}$ is a d_{in}-dimensional vector filled with ones. Note that each output location $\mathbf{i}$ is obtained as the discrete convolution of input pixels at locations $\mathbf{i} + \mathbf{p} - \mathbf{1}$ with filter pixels at location $\mathbf{p}$. This is the result of the inner sum for a given input at depth j. The filter is applied through the depth of the inputs, so that $y(\mathbf{i})$ is the sum of the convolution along the depth.

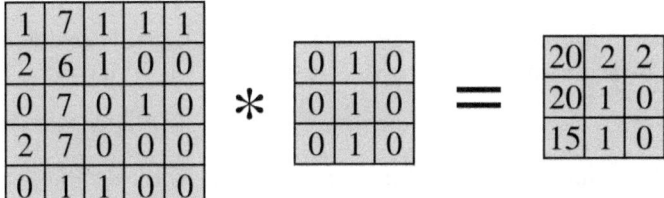

Figure 8.2: A convolution is applied between a 5 × 5 pixel image and a 3 × 3 filter that detects vertical lines. The result of the convolution shows that a vertical line is present on the left of the image.

It is common that a convolution layer has many filters $g^1, g^2, \ldots, g^\ell \in (\mathbb{R}^{(n_{\mathrm{fil}})^{d_{\mathrm{in}}}})^{k_{\mathrm{in}}}$. Then, the output of the convolution layer is $y \in (\mathbb{R}^{(n_{\mathrm{in}}-n_{\mathrm{fil}}+1)^{d_{\mathrm{in}}}})^\ell$, that is, the depth of the output is ℓ.

- Pooling layer: its role is to downsize the objects to reduce n_{in}, by downsampling and selecting the "dominant" pixel within regions of the input. Intuitively, some features of images, say, do not need extremely high resolution to detect patterns, meaning that less pixels are needed. For example, images that differ from a slight shift yield similar downsampled images.

Let n_{pool} be a divider of n_{in}, called the tiling size. A pooling layer partitions the $(n_{\mathrm{in}})^{d_{\mathrm{in}}}$ pixels of the inputs into $(n_{\mathrm{in}}/n_{\mathrm{pool}})^{d_{\mathrm{in}}}$ d_{in}-dimensional squares of size $(n_{\mathrm{pool}})^d$, then maps each square into a single pixel. There are two standard ways of merging pixels defining different pooling layers: *max pooling* returns the maximum value within each square, and *average pooling* returns the mean value within each square. Mathematically, for $\mathbf{i} \in \{1, \ldots, n_{\text{in}}/n_{\text{pool}}\}^{d_{\text{in}}}$ and $j \leq k_{\text{in}}$, this corresponds to

$$y_j(\mathbf{i}) = \max_{p_1, \ldots, p_{d_{\text{in}}} \leq n_{\text{pool}}} x_j(\mathbf{i} + \mathbf{p} - \mathbf{1})$$

for a max pooling layer. For an average pooling layer, we simply replace the max by an average. See Figure 8.3 for a simple example of pooling layer.

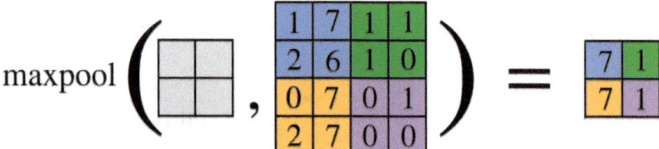

Figure 8.3: A 2×2 max pooling is applied to a 4×4 image, returning a 2×2 matrix with largest pixel value per element of the partition.

- Fully connected layer: equivalent to a layer of a fully connected network, it flattens the input and maps it to the desired output size (e. g., multiclass classification).
- Output activation function: softmax can be applied to obtain a probability vector whose coordinate i corresponds to the predicted probability that an input belongs to class i.

Our definition of convolutional networks is not as general as it can be. Variations are possible, e. g., by padding an image with zeroes at the boundary to preserve the size n through a convolution layer, applying filters every other pixel, etc. The overall structure remains the same: trainable parameters are the values in filters, so that the number of parameters to train in a given convolutional layer with hyperparameters n_{in}, d_{in}, k_{in} for inputs and n_{fil}, k_{in} for the size and number of filters is given by

$$\#\text{filters} \times \#\text{pixels/filter} = k_{\text{in}} \times n_{\text{fil}}^{d_{\text{in}}}.$$

Another important property of convolutional networks is that they are translation invariant by design: if a filter is trained to detect a capybara's ear, then since it is applied identically at every location of an image, it detects a capybara's ear regardless of its location on the image.

8.4.2 Generative adversarial neural networks

The original generative adversarial neural network (GAN) was introduced as a new generative framework from training data sets. It addressed the question: "if you are given a data set of objects with a certain degree of consistency, then can we artificially generate similar objects?"

Mathematically, we are given a dataset $S = \{(x_i, y_i); i = 1, \ldots, m\} \subset \mathcal{X} \times \mathcal{Y}$ with i. i. d. samples generated from a common probability distribution D. The goal of a generative model f is the following: given a label $y \in \mathcal{Y}$, it generates $f(y) = x \in \mathcal{X}$ such that $x \sim D(x|y)$, that is, x is a random variable with conditional distribution $D(\cdot|y)$ given the value of y.

For example, if S is a set of pictures x_is with labels in $\mathcal{Y} = \{\text{"capybara"}, \text{"beaver"}\}$, then a generative model trained on S tries to output a picture $f(\text{"capybara"})$ with same law as the pictures of capybaras in the dataset S.

Two issues rapidly emerge from trying to formally define a generative model:

I: How to avoid that the generative model simply outputs images from the dataset? Roughly speaking, we would like the generative model to "create new images" and not merely memorize previously seen images.

II: What loss function can we train the model on? For regression or classification, the model updates its parameters according to its performance measured using the labels of the dataset. Here, when the model outputs a picture, we would like the model to assess its performance autonomously, i. e., without needing human feedback.

To simplify the discussion, let us forget about the label y and reduce the problem to its equivalent formulation with a dataset $S = \{x_i; i = 1, \ldots, m\} \sim D^m$.

I:
To generate random outputs with a distribution close to the unknown D, we start from a simple, dummy distribution γ on some lower-dimensional space $(d < n)$; for example, we can take γ a standard Gaussian in $\mathbb{R}^d$ with covariance matrix I_n. Typically, to avoid sampling the same value twice (with probability 1), γ is chosen to be absolutely continuous with respect to the Lebesgue measure. Then a random signal $z \overset{d}{\sim} \gamma$ is given to the generative model $G(z) \in \mathbb{R}^n$, whose goal is to map the distribution γ to D, that is, for a perfect generative model, $G(z) \overset{d}{\sim} D$. This leads us to the second issue.

II:
To evaluate the output $G(z)$, we introduce another function f_{disc}, called the discriminator function, which is a classifier whose task is to predict whether a particular sample x is sampled from the true, unknown distribution D or from the generative model G, namely, if $f_{\mathrm{disc}}(x)$ is high, then a good discriminator should classify x as sampled from D, and if $f_{\mathrm{disc}}(x)$ is low, then it should classify x as sampled from G.

In the GAN setup, both G and f_{disc} are neural networks, with their own set of parameters to be trained. The true loss function is given by

$$\mathcal{L}(G, f_{\mathrm{disc}}) := \mathbb{E}_{x \sim D}[\log f_{\mathrm{disc}}(x)] + \mathbb{E}_{z \sim \gamma}[\log(1 - f_{\mathrm{disc}}(G(z)))].$$

The GAN (and its discriminator) aims at solving the minimax problem

$$\min_{G} \max_{f_{\mathrm{disc}}} \mathcal{L}(G, f_{\mathrm{disc}}).$$

This corresponds to a zero-sum game with two players. If G and f_{disc} spanned the whole space of distributions (which is not the case here), we would have $\min_{G} \max_{f_{\mathrm{disc}}} \mathcal{L} = \max_{f_{\mathrm{disc}}} \min_{G} \mathcal{L}$, so the game has a unique optimal solution; see, e. g., the original paper [24], where GANs were introduced. For G and f_{disc} neural networks, it is generally not the case.

In the loss $\mathcal{L}$, for a fixed generator G, $\max_{f_{\mathrm{disc}}} \mathcal{L}(G, f_{\mathrm{disc}})$ optimizes the discriminator to reject samples $G(z)$ that look dissimilar to what f_{disc} knows of the distribution D. The loss is then minimized over generators that fool the best discriminator.

Since we actually cannot compute the first expectation in $\mathcal{L}(G, f_{\mathrm{disc}})$, we approximate it using the dataset $S = \{x_i\}_{i=1}^{m}$:

$$\mathbb{E}_{x \sim D}[\log f_{\mathrm{disc}}(x)] \approx \frac{1}{m_1} \sum_{i=1}^{m} \log f_{\mathrm{disc}}(x_i),$$

whereas the second expectation can either be computed or approximated similarly using $z_1, \ldots, z_m \overset{\text{i.i.d.}}{\sim} \gamma$ as

$$\mathbb{E}_{z \sim \gamma}[\log(1 - f_{\mathrm{disc}}(G(z)))] \approx \frac{1}{m'} \sum_{i=1}^{m'} \log(1 - f_{\mathrm{disc}}(G(z_i))).$$

The optimization is done in two steps: first, update the discriminator f_{disc} taking the gradient with respect to parameters of f_{disc} and computing the gradient ascent update; then update the generator G taking the gradient with respect to parameters of G and computing the gradient descent update.

GANs have become very powerful tools for generative models, where we want to generate similar samples to a given dataset, for example, new faces, new rooms, etc.[3] Furthermore, GANs can be used to solve problems in which we cannot easily formalize a loss function, where we do not have access to a loss function, or we are unable to compute gradients on the loss function.

8.5 Some elements of modern theory of neural networks

Since major progress is made every year in the theoretical understanding of neural networks, this section cannot do justice to the advances of the field. The complexity of neu-

---

3 Check out https://thisxdoesnotexist.com/ for many examples of GANs used to generate a variety of things!

ral networks does not allow for a unifying theory; nonetheless, we present in this section some modern results that we chose for the following reasons: they are mathematically rigorous and elegant but also have practical implications. On the other hand, these results have influenced part of directions that researchers followed, which makes them relevant for the reader interested in present and future developments in deep learning theory.

Neural networks are often used largely overparameterized. As a consequence, for a given architecture, there may be many neural networks that perfectly fit a given dataset. Yet, neural networks trained with (variations of) gradient descent sometimes seem to generalize well or, at least, better than predicted by the bias-variance tradeoff (see Chapter 4), for example, when compared to overparameterized linear regression.

Constructing a general theory to explain their success seems out of reach. In this section, we focus on **overparameterized** neural networks, specifically, on their **initialization** and **training**. We will see that different training regimes occur by only changing the scale of the weights at initialization.

8.5.1 Initialization scale

To simplify the notation, we consider a neural network f_w with **scalar input and output**. Suppose moreover that it has a single hidden layer of width $d \in \mathbb{N}$ with the identity activation function between the hidden and output layers; that is, for an input $x \in \mathbb{R}$, the network predicts

$$f_w(x) := \frac{1}{d^\gamma} w_2 \sigma(w_1 x + b),$$

where $w_1, b \in \mathbb{R}^{d \times 1}$ and $w_2 \in \mathbb{R}^{1 \times d}$ are the learnable parameters of the network, and $\gamma > 0$ is a hyperparameter. We assume that the components of w_1, w_2, and b are i.i.d. standard Gaussian variables $\mathcal{N}(0,1)$ at initialization, so that γ governs the initialization scale (recall that if $X \sim \mathcal{N}(0,1)$, then $aX \sim \mathcal{N}(0, a^2)$ for $a \in \mathbb{R}$).

Instead of the above equation, we will prefer to (equivalently) write the network as the sum of its hidden neurons, that is,

$$f_w(x) := \frac{1}{d^\gamma} \sum_{k=1}^{d} w_{2,k} \sigma(w_{1,k} x + b_k). \tag{8.5}$$

In the Bayesian learning paradigm, the prior distribution of a model (initialization) is crucial as it is expected to encode our prior beliefs and radically influences the learning (training). It is not clear that neural networks trained by gradient descent perform Bayesian learning (in fact, they do not but are somehow related in some cases; see [27]). Nonetheless, Neal [35] observed the following result for 2-layer neural networks (later generalized to deeper networks).

Theorem 8.5.1. *Suppose that f_w is a neural network at initialization defined as in (8.5) with $\gamma = 1/2$. Suppose moreover that σ is a Lipschitz map. Then the law of f_w converges as $d \to \infty$ to $GP(0, K^{(2)})$, where the kernel $K^{(2)}$ is given by*

$$K^{(2)}(x, x') := \mathbb{E}_{g \sim GP(0, K^{(1)})}[\sigma(g(x))\sigma(g(x'))],$$

where $\quad K^{(1)}(x, x') := xx' + 1.$

Sketch of proof. The main idea is to use the central limit theorem (Theorem 2.8.2). Indeed, note that in (8.5) the output $f_w(x)$ corresponds to a sum of d i. i. d. variables and the convergence to a Gaussian distribution follows from the central limit theorem. (We need to make sure that $\sigma(X)$ has a second moment for a Gaussian variable X, which is the case since σ is Lipschitz.)

For the covariance, we first observe that $x \mapsto w_1 x + b$ is a d-dimensional Gaussian process, since for all x, we have $w_1 x \sim \mathcal{N}(\vec{0}, x^2 \mathcal{I}_d)$ independent from $b \sim \mathcal{N}(\vec{0}, \mathcal{I}_d)$, so that $w_1 x + b$ is a centered Gaussian random variable, and the covariance kernel is easily computed as

$$\mathbb{E}[(w_1 x + b)(w_1 x' + b)]$$

$$= \mathbb{E}\left[\sum_{i=1}^{d} w_{1,i}^2 xx'\right] + \mathbb{E}\left[\sum_{i=1}^{d} w_{1,i} x b_i + \sum_{i=1}^{d} w_{1,i} x' b_i\right] + \mathbb{E}\left[\sum_{i=1}^{d} b_i^2\right]$$

$$= 1 \times xx' + \mathbb{E}[w_1]\mathbb{E}[b](x + x') + 1$$

$$= xx' + 1$$

$$= K^{(1)}(x, x').$$

Then we condition on the values of the neurons in the hidden layer $\{x_k^{(1)}; k = 1, \ldots, d\}$ and check that $f_w(x)$ given $x^{(1)}$ is a Gaussian process (as a linear combination of independent Gaussian processes) with covariance kernel $\frac{1}{d}\sum_{k=1}^{d} x_k^{(1)}(x')_k^{(1)}$. We get a limit as $d \to \infty$ that depends neither on $x^{(1)}$ nor on $(x')^{(1)}$, and getting rid of the conditioning yields the claim. $\quad\square$

Knowing that infinitely wide neural networks with the above initialization are Gaussian processes is informative about the prior knowledge we inject in our model but does not provide any insight on what happens during training. In the next two sections, we present two recent techniques that allow us to say more about training.

8.5.2 The neural tangent kernel

In this section, we consider a neural network defined as in (8.5) with $\gamma = 1/2$. Since we study training, we denote by w_t the network parameters at time t of training.

From Theorem 8.5.1 we know that the network at initialization f_{W_0} is, for large width, close to a Gaussian process with explicit kernel. Recall that gradient descent steps are given by

$$W_{t+1} = W_t - \eta \frac{1}{m} \sum_{i=1}^{m} \partial_{\hat{y}} \ell(y_i, \hat{y})|_{\hat{y}=f_{W_t}(x_i)} \nabla_W f_{W_t}(x_i), \tag{8.6}$$

where $\eta > 0$ is the learning rate. We are interested in the dynamics of training, meaning that we would like to know more about how f_{W_t} evolves in time (and in particular what it looks like at convergence as $t \to \infty$ (provided that it converges)). By a first-order Taylor expansion of $f_{W_{t+1}}$ we have

$$f_{W_{t+1}}(x) \approx f_{W_t}(x) + (W_{t+1} - W_t) \cdot \nabla_W f_{W_t}(x).$$

Using (8.6), we get

$$f_{W_{t+1}}(x) - f_{W_t}(x) \approx -\eta \frac{1}{m} \sum_{i=1}^{m} \partial_{\hat{y}} \ell(y_i, \hat{y})|_{\hat{y}=f_{W_t}(x_i)} \nabla_W f_{W_t}(x_i) \cdot \nabla_W f_{W_t}(x). \tag{8.7}$$

It turns out that the dot product in the right-hand side defines a time-dependent kernel

$$\Theta_t^{(d)}(x, x') := \nabla_W f_{W_t}(x) \cdot \nabla_W f_{W_t}(x'),$$

where we made the dependency on the width d explicit. The kernel $\Theta_t^{(d)}$ is called the *Neural Tangent Kernel* (NTK) of the neural network at time t. Because the initialization is random, the NTK at initialization $\Theta_0^{(d)}$ is itself a random kernel. Moreover, the fact that its dynamics in time depends on the training makes the exact dynamics of the network f_{W_t} intractable.

However, there is a very important result by Jacot et al. [29], later generalised in many ways (e. g., by Yang [44]). Consider the gradient flow corresponding to the limit of gradient descent (see Section 3.2.2) as $\eta \to 0+$ and the number of steps T/η for fixed $T \in \mathbb{R}_+$ and given by

$$\partial_t f_{W_t}(x) = -\frac{1}{m} \sum_{i=1}^{m} \partial_{\hat{y}} \ell(y_i, \hat{y})|_{\hat{y}=f_{W_t}(x_i)} \Theta_t^{(d)}(x_i, x). \tag{8.8}$$

Theorem 8.5.2. *There exists a deterministic kernel $\Theta^{(\infty)} : \mathbb{R} \times \mathbb{R} \to \mathbb{R}$ such that, in the setup above, for σ Lipschitz, twice differentiable with bounded second derivative, for all $x, x' \in \mathbb{R}$ and $T \in \mathbb{R}_+$, we have with probability 1 that*

$$\lim_{d \to \infty} \sup_{t \leq T} |\Theta_t^{(d)}(x, x') - \Theta^{(\infty)}(x, x')| = 0.$$

Furthermore, the limiting NTK has the following expression:

$$\Theta^{(\infty)}(x,x') = K^{(1)}(x,x')\dot{K}^{(2)}(x,x') + K^{(2)}(x,x'),$$

where $K^{(1)}$ and $K^{(2)}$ are the kernels defined in Theorem 8.5.1, and $\dot{K}^{(2)}$ is defined as $K^{(2)}$ with the derivative σ' instead of σ in its definition.

Remark 8.5.3. The assumption that the neural network has a single hidden layer is superfluous, and we make it solely for presentation.

Remark 8.5.4. We can show that for nonpolynomial, Lipschitz σ, the restriction of the NTK $\Theta^{(\infty)}$ to the unit sphere is positive semidefinite. (In our case – scalar inputs – it does not make much sense; however, it does in the general d-dimensional case.)

Remark 8.5.5. The assumptions on σ can be greatly relaxed, as it suffices to assume that the second-order weak derivative of σ is polynomially bounded. (It is the case, for example, of the ReLU activation.)

Many consequences can be drawn from Theorem 8.5.2; we now loosely discuss the most straightforward.

Training follows kernel gradient descent

With kernel gradient descent (8.8), taking the limit as $d \to \infty$, the NTK is fixed, and the training dynamics is given by

$$\partial_t f_{w_t} = \frac{1}{m} \sum_{i=1}^{m} \partial_{\hat{y}} \ell(y_i, \hat{y})|_{\hat{y}=f_{w_t}(x_i)} \Theta^{(\infty)}(x_i, x).$$

Convergence

Suppose that the loss is the mean square loss $\ell(y, y') = \frac{1}{2}(y - y')^2$. The training trajectory becomes (up to a factor 2)

$$\partial_t f_{w_t}(x) = -\frac{1}{m} \sum_{j=1}^{m} \Theta^{(\infty)}(x, x_j)(f_{w_t}(x_j) - y_j).$$

Suppose that $(\Theta^{(\infty)}(x_i, x_j))_{1 \le i,j \le m}$ is positive semidefinite. This differential equation has an explicit solution, and in particular it can be shown that f_{w_t} converges as $t \to \infty$ to

$$f_{w_\infty}(x) = f_{w_0}(x) - \sum_{i,j=1}^{m} \Theta^{(\infty)}(x, x_i)\Theta^{(\infty),-1}(x_i, x_j)(f_{w_0}(x_j) - y_j), \tag{8.9}$$

where $\Theta^{(\infty),-1}$ is the inverse of $\Theta^{(\infty)}$. We see that choosing x in the dataset, i. e., some x_i, we have $f_{w_\infty}(x_i) = y_i$ so that the network perfectly fits the data, and the empirical loss is zero.

Gaussian process

Since f_{W_0} is a Gaussian process by Theorem 8.5.1 and since we apply a linear map to the Gaussian vector $(f_{W_0}(x_j))_{j=1,\dots,m}$ in the expression of f_{W_∞} in (8.9), the limit neural network f_{W_∞} is itself a Gaussian process.

It is worth noting that this Gaussian process does not correspond to the Bayesian posterior given the prior f_{W_0} unless we subtract the initial output function f_{W_0}; see, e.g., [27] for more detail.

Some limitations

The NTK regime (also called the *kernel regime* or *lazy regime*) does not fully describe neural network behavior for several reasons. The first one is that neural networks used in practice do not contain an infinite number of neurons. Nonetheless, this shows that overparameterization is not always problematic and is sometimes desirable: firstly, the fact that a global minimum is found by gradient descent is not trivial since the problem is nonlinear. Secondly, we can see that the trained network on unseen datapoints $f_{W_\infty}(x)$ in (8.9) does not explode and its variance can be explicitly computed. Another critic that can be made to the NTK regime is that the kernel is fixed as a result of the fact that individual weights asymptotically do not move during training: the map f_{W_t} evolves during training as a result of infinitely many infinitesimal changes in the weights of the neurons. Loosely speaking, this causes the network to not learn features to fit the data, akin to a kernel method with given kernel, which fits the data as a linear combination of feature maps of the kernel, without trying to learn these features at any point. However, feature learning can be a crucial characteristic of successful models in practice; see, e. g., [15] for a language representation model using pretraining to learn important features of the data.

8.5.3 Mean field regime

In this section, we stay very informal to present another training regime that allows us to obtain theoretical guarantees.

We now assume that our 2-layer neural network is initialized as in (8.5) with $\gamma = 1$. Compared to the previous section, where we supposed $\gamma = 1/2$, this initialization is smaller. Let μ_0 be a probability distribution on $\mathbb{R}^3$, seen as the law of the weights of a single neuron at initialization (they are not Gaussian anymore). More specifically, suppose that i. i. d. $(w_{1,k}, w_{2,k}, b_k)$ are sampled from μ_0, and let

$$\mu^d = \frac{1}{d} \sum_{k=1}^d \delta_{(w_{1,k}, w_{2,k}, b_k)}.$$

We assume moreover that the components w_1, w_2, and b are independent under μ_0. The above is a probability measure on the weight space $\mathbb{R}^3$ consisting of the average of Dirac masses on the neuron weights of the neural network at initialization. The parameterization (8.5) with $\gamma = 1$ can now be rewritten in the integral form

$$f_{\mu_0^d}(x) = \int_{\mathbb{R}^3} w_2 \sigma(w_1 x + b) \mu_0^d(dw_1, dw_2, db),$$

where $f_{\mu_0^d}$ is the network at initialization. Suppose that the weights have expectation 0, and note that $\lim_{d \to \infty} f_{w_0^d} \equiv 0$ almost surely by the law of large numbers:

$$\lim_{d \to \infty} f_{\mu_0^d}(x) = \lim_{d \to \infty} \int_{\mathbb{R}^3} w_2 \sigma(w_1 x + b) \mu_0^d(dw_1, dw_2, db)$$

$$= \lim_{d \to \infty} \sum_{k=1}^{d} w_{2,k} \sigma(w_{1,k} x + b_k)$$

$$= \int_{\mathbb{R}^3} w_2 \sigma(w_1 x + b) \mu_0(dw_1, dw_2, b)$$

$$= 0,$$

where we first integrated over w_2 to get 0 above (thanks to the independence of w_1, w_2, and b).

We denote by μ_t^d the measure of the neural network initialized at μ_0^d at time t of training. The advantage of writing the neural network in integral form and taking the limit is that the dynamics of the measure μ_t^d during training can be studied instead of that of $f_{\mu_t^d}$. Informally, we can make sense of $\lim_{d \to \infty} \mu_t^d = \mu_t$, where μ_t corresponds to the infinite-width neural network at time t of training and initialized at μ_0. As many tools that are out of the scope of this book are involved (from optimal transport theory such as Wasserstein gradient flow), we refrain to go into technical details and refer the interesting reader to [10]. One of the main results of this paper can be informally stated as follows.

Theorem 8.5.6 (Informal). *Under some assumption on the support of the initial measure μ_0 and choosing $\sigma = $ ReLU, if μ_t converges to some μ_∞ as $t \to \infty$, then μ_∞ is optimal in the sense that the induced network f_{w_∞} achieves zero loss.*

Remark 8.5.7. Such a result is called a *global convergence result*: it does not claim that the network converges, but if it does, then its limit perfectly fits the data. It is important to note that the results in the mean-field regime require the assumption of having a single hidden layer (on top of infinite width). We chose $\sigma = $ ReLU, but the result holds for more general activations under some homogeneity assumption of the network.

Remark 8.5.8. As opposed to the NTK regime, in the mean-field regime, feature learning occurs, and we can move from one to another regime simply by changing the initialization scale. For details on the impact of initialization scale on feature learning, see [45].

8.5.4 Edge of chaos

We discussed the problem of exploding and vanishing gradients in Section 8.2.2 and explained how a heuristic justified commonly used weight initialization schemes in Section 8.2.3. The idea is to keep the magnitude of the propagated signal variance constant across layers, both forward and backward. A more complete but similar argument that we present now looks at the whole covariance kernels of the hidden layers. It leads to more precise recommendations regarding initialization, which practitioners are not always aware of.

Recall the recursive definition (8.1) of neural networks using preactivations and postactivations. Suppose that the data are centered with variance 1. Consider a deep neural network with a differentiable activation function σ and random parameters such that weights are independent centered Gaussians with variance $\frac{v_w}{d_\ell} > 0$ at layer $\ell = 1, \ldots, L$ and biases are independent Gaussians with variance $v_b \geq 0$. Let

$$\mathbf{K}(x, x'; K) := \mathbb{E}_{g \sim \mathrm{GP}(0, K)}[\sigma(g(x))\sigma(g(x'))],$$

where K is a positive definite kernel, and $x, x' \in \mathbb{R}^{d_0}$. Extending Theorem 8.5.1 to deep networks, we claim that the covariance kernels of the preactivations $\tilde{a}^{(\ell)}$ for large widths are close to

$$K^{(1)}(x, x') = v_w x x' + v_b,$$
$$K^{(\ell+1)}(x, x')_{ij} = \delta_{ij} v_w \mathbf{K}(x, x'; K^{(\ell)}) + \delta_{ij} v_b,$$

where $1 \leq i, j \leq d_\ell$ are the indices of the neurons at layer ℓ. This means that $\tilde{a}_i^{(\ell)}$ is a Gaussian process with covariance kernel $K^{(\ell)}$ for all i and is independent of $\tilde{a}_j^{(\ell)}$ for all $j \neq i$. Indeed, if $\tilde{a}^{(\ell-1)}$ is a Gaussian process with covariance kernel $K^{(\ell-1)}$, then by independence of $\tilde{a}^{(\ell-1)}$ with $W^{(\ell+1)}$ and $b^{(\ell+1)}$, we have for all $1 \leq i, j \leq d_\ell$ that

$$\mathbb{E}[\tilde{a}_i^{(\ell)}(x)\tilde{a}_j^{(\ell)}(x') | \tilde{a}^{(\ell-1)}(x), \tilde{a}^{(\ell-1)}(x')]$$
$$= \mathbb{E}[W_{i:}^{(\ell)}\sigma(\tilde{a}^{(\ell-1)}(x))W_{j:}^{(\ell)}\sigma(\tilde{a}^{(\ell-1)}(x')) | \tilde{a}^{(\ell-1)}(x), \tilde{a}^{(\ell-1)}(x')] + \mathbb{E}[b_i^{(\ell)}b_j^{(\ell)}]$$
$$= \sum_{k,n=1}^{d_{\ell-1}} \mathbb{E}[W_{ik}^{(\ell)}W_{jn}^{(\ell)}]\sigma(\tilde{a}_k^{(\ell-1)}(x))\sigma(\tilde{a}_n^{(\ell-1)}(x')) + \delta_{ij}v_b$$
$$= \sum_{k=1}^{d_{\ell-1}} \delta_{ij}\mathbb{E}[W_{ik}^{(\ell)}W_{jk}^{(\ell)}]\sigma(\tilde{a}_k^{(\ell-1)}(x))\sigma(\tilde{a}_k^{(\ell-1)}(x')) + \delta_{ij}v_b$$
$$= \delta_{ij}\frac{v_w}{d_{\ell-1}} \sum_{k=1}^{d_{\ell-1}} \sigma(\tilde{a}_k^{(\ell-1)}(x))\sigma(\tilde{a}_k^{(\ell-1)}(x')) + \delta_{ij}v_b.$$

Taking the expectation on both sides shows that $\mathbb{E}[\tilde{a}_i^{(\ell)}(x)\tilde{a}_j^{(\ell)}(x')] = K^{(\ell)}(x, x')_{ij}$, showing that hidden neurons at a given layer are independent from each other and such that for each neuron $i \leq d_\ell$, the covariance of $\tilde{a}_i^{(\ell)}(x)$ and $\tilde{a}_i^{(\ell)}(x')$ is $K^{(\ell)}(x, x')$, where we henceforth write $K^{(\ell)}(x, x')$ for $K^{(\ell)}(x, x')_{ii}$, which does not depend on i.

Let us make some simplifying assumptions: from the equation above, it is natural to take $v_w + v_b = 1$ and rescale σ so that $\mathbb{E}_{X \sim \mathcal{N}(0,1)}[\sigma(X)^2] = 1$ for all x with norm 1. Indeed, this ensures that the variance $K^{(\ell)}(x, x) = 1$ for all $x \in \mathbb{R}^{d_0}$ such that $\|x\| = 1$. Henceforth, we thus write $1 - v_w$ in place of v_b and assume that $\mathbb{E}_{X \sim \mathcal{N}(0,1)}[\sigma(X)^2] = 1$.

For $\rho \in [-1, 1]$, define the matrix

$$C(\rho) := \begin{pmatrix} 1 & \rho \\ \rho & 1 \end{pmatrix}.$$

We define $\hat{\sigma}$, the *dual activation function*[4] of σ, by

$$\hat{\sigma}(\rho) := \mathbb{E}_{(Z_1, Z_2) \sim \mathcal{N}(0, C(\rho))}[\sigma(Z_1)\sigma(Z_2)].$$

Finally, let $h : \rho \mapsto v_w \rho + 1 - v_w$. The preactivation kernel recursion can be rewritten as

$$K^{(\ell+1)}(x, x') = (h \circ \hat{\sigma})^{\circ(\ell)} h(K^{(1)}(x, x')).$$

Under mild assumptions, it is standard that such repeated compositions converge to a fixed point, as the number of compositions goes to infinity, by the Banach fixed point theorem. They do here, and we admit the following properties of the dual activation function (for the proofs, see [13]): if $\mathbb{E}_{X \sim \mathcal{N}(0,1)}[\sigma(X)^2] = 1$, then
1. $\hat{\sigma}$ is convex.
2. $\hat{\sigma}' = \widehat{\sigma'}$, that is, the derivative of the dual is the dual of the derivative.
3. $\hat{\sigma}(\rho) > \rho$ for all $\rho \in (-1, 0)$.

Since h is an increasing linear map, the composition $h \circ \hat{\sigma}$ itself is convex. Note that the last property entails that there cannot be any fixed point in $(-1, 0)$. We also have that

$$(h \circ \hat{\sigma})(0) = h(\mathbb{E}_{X \sim \mathcal{N}(0,1)}[\sigma(X)]^2) \in (0, 1);$$
$$(h \circ \hat{\sigma})(1) = h(\mathbb{E}_{X \sim \mathcal{N}(0,1)}[\sigma(X)^2]) = 1.$$

By the convexity the existence of a fixed point in $[0, 1)$ depends on the sign of $(h \circ \hat{\sigma})'(1) - 1$: from Figure 8.4 we have that
1. If $(h \circ \hat{\sigma})'(1) < 1$, then there is no fixed point in $[0, 1)$.
2. If $(h \circ \hat{\sigma})'(1) > 1$, then there is a unique fixed point in $[0, 1)$.

In the case $(h \circ \hat{\sigma})'(1) = 1$, infinitely many fixed points can exist if $h \circ \hat{\sigma}$ is the identity. The condition can be made more explicit by computing

$$r_\sigma(v_w) := (h \circ \hat{\sigma})'(1) = v_w \hat{\sigma}'(1) = v_w \widehat{\sigma'}(1) = v_w \mathbb{E}_{X \sim \mathcal{N}(0,1)}[\sigma'(X)^2].$$

---

4 The term is taken from [13].

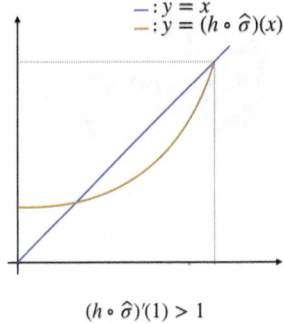

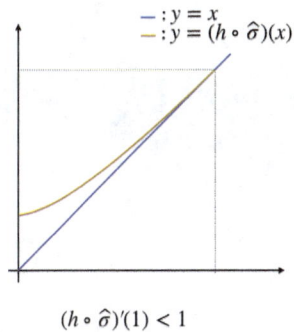

Figure 8.4: Fixed points of $h \circ \hat{\sigma}$ in $[0, 1)$. **Left panel:** $(h \circ \hat{\sigma})'(1) > 1$, so the graph of $h \circ \hat{\sigma}$ intersects the line $y = x$ once for $0 \le x < 1$, and this point is unique by convexity. **Right panel:** $(h \circ \hat{\sigma})'(1) < 1$, so by convexity $(h \circ \hat{\sigma})(x) > x$ for all $x \in [0, 1)$.

This shows that as the depth ℓ grows large, the activation kernel $K^{(\ell)}$ exhibits two phases:

1. **Order** $r_\sigma(v_w) < 1$: the covariance $\Sigma^{(\ell)}(x, x')$ converges to 1 for all inputs $x, x' \in \mathbb{R}^{d_0}$ with norm 1.

2. **Chaos** $r_\sigma(v_w) > 1$: the covariance $\Sigma^{(\ell)}(x, x')$ converges to $c \in [0, 1)$ for all inputs $x \ne x' \in \mathbb{R}^{d_0}$ with norm 1.

In the ordered phase, any structure in the data is not seen by the neural network, whereas in the chaotic phase, two points $x \ne x'$ do not have their correlation at layer ℓ converge to 1 even when x and x' are very close to each other (but distinct), indicating that the neural network output is very rough. It has thus been advocated to initialize the weights of the neural network with the variance parameter v_w on *the edge of chaos* $r_\sigma(v_w) = 1$ or very close to 1. See, e. g., [37, 26, 43].

The phase diagram of the order/chaos dichotomy is more general than what we have presented here: we assumed that $v_b = 1 - v_w$, which is not necessary, in which case the edge of chaos $r_\sigma(v_w, v_b) = 1$ is described as a curve in a two-dimensional plane. The general edge of chaos curves $r(v_w, v_b) = 1$ are slightly more cumbersome to characterize, and we refer to [26] for details on how to obtain them; see Figure 8.5 for plots of three order/chaos phase diagrams.

Lastly, we note that the variances $(v_b, v_w) = (0, 1)$ of Xavier initialization (init i) are at the edge of chaos for tanh activation, which is also the case of He initialization (init iii), using $(v_b, v_w) = (0, 2)$ for ReLU activation.

8.6 Practical implementation considerations

This section is loosely based on the lecture series [34], advice that has come from many years of theory and experimentation, which have lead to substantial differences in terms of speed, ease of implementation, and accuracy when it comes to putting algorithms to

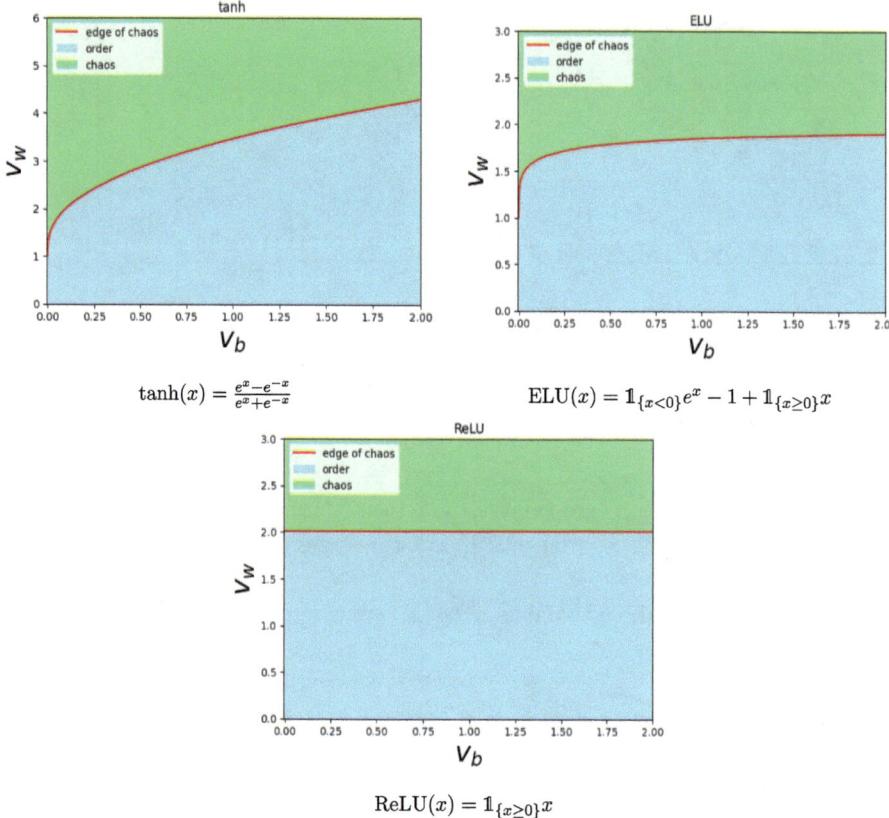

$$\tanh(x) = \frac{e^x - e^{-x}}{e^x + e^{-x}}$$

$$\mathrm{ELU}(x) = \mathbb{1}_{\{x<0\}}e^x - 1 + \mathbb{1}_{\{x\geq0\}}x$$

$$\mathrm{ReLU}(x) = \mathbb{1}_{\{x\geq0\}}x$$

Figure 8.5: Order/chaos phase diagram for three common activation functions for $x \in [0, 2]$. For a given activation function, if the chosen variances (v_w, v_b) are located below the corresponding edge of chaos curve (blue area), then the network is in the chaotic regime, and if they are above the edge of chaos curve (green area), then they are in the order regime.

work in practice. A lot of the information on these lecture series is already implemented (by default) on machine learning libraries (e. g., tensorflow, keras, torch).

In this section, we give a summary and justification/reasoning to why these "tricks of the trade" are important in practice. Many of the advices are quite general and apply to most neural networks.

What things can go wrong when using neural network models?[5]

- Poor performance (overfitting, stuck in local minima, etc.).
- Inability to train network (exploding gradients (NaN values) and flat gradients (see Section 8.2.2), diverging solutions, slow convergence).
- Very large hyperparameter space.

---

5 This is of course not an exclusive problem of neural networks.

8.6.1 Input regularization

We have seen in Chapter 5 that sometimes it is necessary to normalize the input (linear regression with regularization). An input X is typically centered by subtracting the empirical mean $\hat{X}$ and dividing by the empirical variance σ_X^2:

$$X' = \frac{X - \hat{X}}{\sigma_X^2}.$$

For neural networks, normalizing the input is also a common practice, namely:
- Average of each input over the training set should be close to zero.
- Scale input variables so that their covariances are about the same magnitude.
- Input variables should be uncorrelated if possible.

Shifting and scaling is quite simple. However, decorrelating the inputs may be tricky. Sometimes, *principal component analysis* is applied to the feature matrix to remove linear correlations in the input.

It is also observed that convergence is usually faster if the average of each input over the training set is close to zero; in particular, there are both positive and negative values. Let us consider an example of the extreme case.

Example 8.6.1. Suppose all the components of an input vector are positive (the discussion is equivalent when they are all negative). The gradient update for any weight i, j that sits on layer n, w_{ij}^n, is given as

$$w_{ij}^{t+1} = w_{ij}^t - \eta \frac{\partial L}{\partial w_{ij}}, \quad \text{where} \quad \frac{\partial L}{\partial w_{ij}^n} = x_j^{n-1} \frac{\partial L}{\partial z_i^n}.$$

Let us focus on the first layer, $n = 1$. The update for the weights corresponding to a particular output node i (corresponding to row i of the weight matrix W^1) are proportional to $x_j^0 \frac{\partial L}{\partial z_i^1}$. If all components of an input vector are positive, then all updates of the weights that feed into node i will have the same sign (sign($\frac{\partial L}{\partial z_i^1}$)). This means that those weights can only all decrease or increase together for a given input. If the weight vector must change the direction, it can only do so by zigzagging, which is inefficient and slow.

Normalizing the inputs is not only a concern for the speed of convergence but also for the trainability of the network. Due to finite numerical precision, numerical overflow can occur, turning gradient updates into *NaN* updates.

8.6.2 Stabilizing the gradients

We saw in Section 8.2.3 the problems of vanishing and exploding gradients when training a neural network. We discussed how avoiding these problems motivated different weight initialization schemes. We have, however, no guarantee that these initializations prevent these two phenomena to occur beyond the first step of gradient descent. We present here two techniques that have been introduced to keep the training stable.

Gradient clipping

What can we do if the gradient becomes too large? Simply normalize it! The following technique is called *gradient clipping* and is quite intuitive: For $\epsilon > 0$,
- If $\|\nabla_w C(w)\| > \epsilon$, then update the parameters with $-\epsilon \frac{\nabla_w C(w)}{\|\nabla_w C(w)\|}$.
- Otherwise, use the usual update $-\nabla_w C(w)$.

If some components of the gradient dominate, then the others, after clipping, might end up too small. Another practice is to clip each component of the gradient independently if it crosses a threshold ϵ.

Batch normalization

Batch normalization is a way to guarantee that all activations of layers are centered and normalized with respect to the batch of data samples the network is training on. We can formally define it as a new type of layer of the neural network. A batch normalization layer, denoted by BN, does the following:
- $\{x^{(i)}; i = 1, \ldots, m\} \subset \mathbb{R}^d$ is some batch of inputs.
- $\mu := \frac{1}{d} \sum_{i=1}^{d} x^{(i)}$ is the empirical mean (vector) of the input batch.
- $v := \frac{1}{d} \sum_{i=1}^{d} (x^{(i)} - \mu)^2$ is the empirical variance (vector) of the input batch, where the square function is applied componentwise.
- $\overline{x}^{(i)} := \frac{x^{(i)} - \mu}{\sqrt{v + \epsilon}}$ for all $i = 1, \ldots, m$, where $\epsilon > 0$ is a hyperparameter that prevents the denominator to be null.
- $\mathrm{BN}(x^{(i)}) := (\gamma_k \overline{x}_k^{(i)} + \beta_k)_{k=1,\ldots,d}$, where γ_k and β_k are parameters that can be **learnable**.

A neural network can then be composed of many layers, some of them being BN. Despite the several heuristics and the empirical success of batch normalization, there is, for now, neither very robust theory nor consensus on the different effects of batch normalization.

8.6.3 Preventing overfitting

Avoiding overfitting is a common desire in machine learning to have better hope that the model will generalize to unseen data. For neural networks, there are different strategies to prevent overfitting.

Regularization

Similarly to previously seen models, we can also add a regularization term on the loss function, penalizing the model parameters through an L_1 or L_2 norm.

Early stopping

The rational about early stopping is that when we decide on the number of epochs to train a network, it is not usually a very well-informed decision. One strategy is to monitor the training and a validation loss, stopping when the validation loss no longer decreases or starts to increase (see Figure 8.6). A typical early stopping strategy is the following:

1. Split the training data into a training set and a validation set, e. g., in a 2-to-1 proportion.
2. Train only on the training set and evaluate the per-example error on the validation set once in a while, e. g., after every fifth epoch.
3. Stop training as soon as the error on the validation set is higher than it was the last time it was checked.
4. Use the weights the network had in that previous step as the result of the training run.

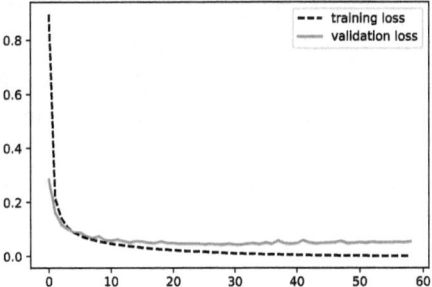

Figure 8.6: Loss curves for the training and validation sets. The early stopping criterion is triggered here because the validation loss is increasing after several epochs.

As we see, choosing a stopping criterion predominantly involves a tradeoff between training time and generalization error. However, some stopping criteria may typically find better tradeoffs than others.

Empirically, it accelerates network training, prevents overfitting, but can be "tricky" to get right due to nonstraightforward behavior of the errors and the presence of tunable parameters. For example, when is the validation error too large with respect to the training error? What is the number of successive epochs that we should wait for?

Dropout

Another efficient way of regularizing neural networks and enhancing their generalization performances is to turn off neurons at random while training. This method is called *dropout*. It works as follows at each training epoch:

1. Independently turn off each neuron with probability $1 - p \in (0, 1)$, where p is called the *keep probability*, whereas $1 - p$ is called the *dropout rate*. The keep probability is an additional hyperparameter to the model.

2. The subnetwork consisting of neurons that have been kept is trained for one step of gradient descent (or any optimization algorithm at use). Only the weights of the neurons that have been kept are updated during this step.

After training, to make predictions with the trained neural network – or to evaluate it on a test dataset – the full neural network is used, with weights rescaled by the keep probability p. The reason why we rescale the weights when predicting is because when the weights are updated, the subnetworks used contain an average proportion p of neurons, and hence their weights tend to be greater than they should when using all of them together. For example, think about the sum of two perfect predictors: the obtained predictor performs poorly unless we divide its output by 2.

Why is it a good idea to use dropout? As we said, this method trains smaller networks inside the network, thus encouraging learning more sparse functions (i. e., simpler, needing less parameters to be expressed). Hence each subnetwork has less opportunities to overfit. Another informal reason is that the full network resembles an average of smaller networks. Averaging simple predictors to construct a good predictor can be a very good idea to enhance training and performance; in the next chapter, we will see a model based on that idea, where the simple predictors do not even need to perform that well individually.

Dropout may also refer to other similar methods, where individual weights are dropped out instead of individual neurons.

8.6.4 Dealing with hyperparameters

Architecture hyperparameters define the hypothesis class $\mathcal{H}$ where our approximator lives in. Some hyperparameters are the number of layers, width of each layer, activation functions, other types of layers (dropout, normalizing layers, etc.).

Then we also have *training hyperparameters*, that specify how the training algorithm $\mathcal{A}$ behaves: the number of epochs, batchsize, learning rate (momentum, etc.), training set/validation set split, loss function (type, L1–L2 penalties).

Many times, good hyperparameters come from experience, trial and error, theoretical justifications, or empirical results. There are also computational frameworks that can help explore this large hyperparameter space, such as hyperopt [7].

8.7 Implementation details

In this section, we present some code snippets that use pytorch to implement a fully connected neural network for scalar regression. More examples for different types of neural networks and experiments can be found on github.

Listing 8.1: Creating dataset.

```
 1 import torch
 2 import torch.utils.data as data
 3
 4 # Generate dataset in tensor format
 5 nData = 400
 6 x_train = torch.rand(nData)
 7 y_train = 1-torch.cos(2*torch.pi*x_train)
 8 x_test = torch.rand(50)
 9 y_test = 1-torch.cos(2*torch.pi*x_test)
10
11 # Custom pytorch dataset
12 class CustomDataset(data.Dataset):
13   def __init__(self, x,y):
14           self.x = x
15           self.y = y
16
17   def __len__(self):
18           return len(self.x)
19
20   def __getitem__(self, idx):
21           sample = {'x': self.x[idx], 'y': self.y[idx]}
22           return sample
23
24 batch_size = 4
25 training_data = CustomDataset(x_train,y_train)
26 training_loader = data.DataLoader(training_data,
        batch_size=batch_size, shuffle=True)
```

Listing 8.2: Fully connected neural network.

```
 1 import torch.nn as nn
 2
 3 class Net(nn.Module):
 4     def __init__(self, nInput, nOutput):
 5         super(Net, self).__init__()
 6         self.linear_relu_stack = nn.Sequential(\
 7                         nn.Linear(nInput, 20),\
 8                         nn.ReLU(),\
 9                         nn.Linear(20, 20),\
10                         nn.ReLU(),\
11                         nn.Linear(20, nOutput))
```

```
12
13      def forward(self, x):
14          return self.linear_relu_stack(x)
```

Listing 8.3: Training and evaluation of the network.

```
 1 def train_one_epoch(training_loader, model, optimizer,
        loss_fn):
 2      running_loss = 0.
 3
 4      for i, data in enumerate(training_loader):
 5          inputs = data['x'].reshape((4,1))
 6          labels = data['y'].reshape((4,1))
 7          optimizer.zero_grad() # set gradient to zero
 8          outputs = model(inputs) # make predictions
 9          loss = loss_fn(outputs, labels) # compute loss
10          loss.backward() # compute gradient
11          optimizer.step() # update learning weights
12          running_loss += loss.item()
13
14      print('Loss: ', running_loss)
15      return running_loss
16
17 model = Net(1,1)
18 loss_fn = torch.nn.MSELoss()
19 optimizer = torch.optim.SGD(model.parameters(), lr=0.01,
        momentum=0.9)
20 scheduler = torch.optim.lr_scheduler.ExponentialLR(
        optimizer, gamma=0.99)
21 n_epochs = 100
22
23 # training for n_epochs
24 for i in range(n_epochs):
25      _ = train_one_epoch(training_loader, model, optimizer,
        loss_fn)
26      scheduler.step()
27
28 with torch.no_grad():
29      model.eval()
30      x_test = x_test.reshape((50,1))
31      y_test_prediction = model(x_test)
32      print("Test loss:", loss_fn(y_test_prediction, y_test)
        )
```

9 Ensemble methods

Ensemble methods are usually reserved for methods that generate a model using an aggregate of *base hypothesis*. There are different ways to approach the construction of ensemble methods. In this chapter, we focus on **boosting**.

Boosting is an algorithmic paradigm that grew out of a theoretical question and became a very practical machine learning tool. The boosting approach uses a generalization of linear predictors to address two major issues:

– The first is the **bias-complexity tradeoff**. We have seen that the generalization error of an ERM learner can be decomposed into a sum of **approximation error** and **estimation error**, as described in equation (4.2). The more expressive the hypothesis class, the smaller the approximation error, but the larger the estimation error can become. A learner is thus faced with the problem of picking a good tradeoff between these two considerations. The boosting paradigm allows the learner to have smooth control over this tradeoff. Learning starts with a **simple class** (which may have a large approximation error), and as it progresses, the class that the predictor may belong to **grows richer**.

– The second issue that boosting addresses is the computational complexity of learning. A boosting algorithm amplifies the accuracy of *weak learners*. Intuitively, we can think of a *weak learner* as an algorithm that uses a simple "rule of thumb" to output a hypothesis that comes from an easy-to-learn hypothesis class and performs just slightly better than a random guess. When a *weak learner* can be implemented efficiently, boosting provides a tool for aggregating such *weak hypotheses* to approximate gradually a good predictor.

Figure 9.1 illustrates how properly aggregating weak hypothesis can solve a task on which a weak hypothesis taken individually does not perform well.

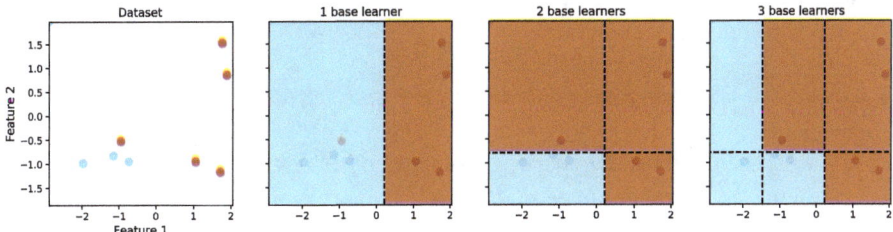

Figure 9.1: An illustration of how a combination of simple base hypothesis behaves on a toy problem.

In this chapter, we start again by considering binary classification for the sake of exposition, though ensemble methods apply to general classification tasks and regression.

https://doi.org/10.1515/9783111288994-009

9.1 Weak learner

Let us first define the concept of γ-weak-learnability. In a binary classification setting, if we were to randomly guess labels by tossing a fair coin each time, then in average, we would be right half of the time and wrong just as much. For $\gamma \in (0, 1/2)$, usually small, a γ-weak-learner is a learner that merely does slightly better than that where it "slightly" depends on γ.

Recall the realisability assumption in Definition 4.1.2 for $\mathcal{H}, \mathcal{D}, f$: there exists $h \in \mathcal{H}$ such that $R_{\mathcal{D},f}(h) = 0$.

Definition 9.1.1. A hypothesis class $\mathcal{H}$ is γ-weak PAC-learnable if there exists a function $m_{\mathcal{H}} : (0,1) \to \mathbb{N}$ with the following property: for every $\delta \in (0,1)$, every distribution $\mathcal{D}$ over $\mathcal{X}$, and every labeling function $f : \mathcal{X} \to \{0,1\}$, when running $\mathcal{A}_{\text{ERM}}$ on $m \geq m_{\mathcal{H}}(\delta)$ i. i. d. samples, the hypothesis $h_{\mathcal{A}}$ is such that with probability at least $1 - \delta$,

$$R_{\mathcal{D}}(h_{\mathcal{A}}) \leq 1/2 - \gamma.$$

Note that this definition is almost identical to the definition of PAC learnability in Definition 4.1.4 with the key difference that PAC learnability implies the ability to find an arbitrarily good classifier with generalization error at most (arbitrarily small) $\epsilon > 0$, whereas in weak learnability, we need to only output a hypothesis with generalization error at most $1/2 - \gamma$ for fixed $\gamma > 0$, that is, whose error rate is slightly better than that given by random labeling. The hope is that it may be easier to come up with efficient weak learners than with efficient (full) PAC learners.

One idea is to take a "simple" hypothesis class, the base class denoted by $\mathcal{B}$, and to apply ERM with respect to $\mathcal{B}$ as the weak learning algorithm, and then find a way to combine these weak learners into a strong learner. Ideally, we want $\text{ERM}_{\mathcal{B}}$ to be efficiently implementable and any $\text{ERM}_{\mathcal{B}}$ hypothesis to have an error of at most $1/2 - \gamma$.

Remark 9.1.2. What can be considered base classes?
- Decision stumps:

$$\mathcal{H}_{DS} = \{x \mapsto (\text{sign}(\theta - x_j)b)_{j=1,\dots,d}; x \in \mathbb{R}^d, \theta \in \mathbb{R}, b \in \{\pm 1\}\}.$$

They partition the feature space in a dimension j using a threshold θ.
- Decision trees: Let the feature space be $\mathbb{R}^d$. A binary regression tree with k-depth is given by

$$h(x) = \sum_{B \in \mathcal{A}} (-1)^{\mathbb{1}_{\{B_{-1}=0\}}} \mathbb{1}_{\{x \in B\}},$$

where $\mathcal{A}$ is the set of sets defined recursively: start with $\mathcal{A}^{(0)} = \{\mathbb{R}^d\}$, then for fixed $\theta^{(0)} \in \mathbb{R}, b^{(0)} \in \{\pm 1\}$, and $j^{(0)} \in \{1, \dots, d\}$, we can generate $\mathcal{A}^{(1)} = \{A_0, A_1\}$ where $A_0 = \{x \in \mathbb{R}^d : (\theta^{(0)} - x_{j^{(0)}})b^{(0)} < 0\}$ and $A_1 = \{x \in \mathbb{R}^d : (\theta^{(0)} - x_{j^{(0)}})b^{(0)} \geq 0\}$. Then the next partition is defined in A_0 or A_1 according to some criteria (e. g., minimizing the ERM), again given fixed $\theta^{(1)} \in \mathbb{R}$, $b^{(1)} \in \{\pm 1\}$, and $j^{(1)} \in \{1, \ldots, d\}$, we attain $\mathcal{A}^{(2)} = \{A_{00}, A_{01}, A_{11}\}$, where $A_{01} = \{x \in A_0 : (\theta^{(1)} - x_{j^{(1)}})b^{(1)} \geq 0\}$, and $A_{00} \cup A_{01} = A_0$ and $A_{00} \cap A_{01} = \emptyset$. By B_{-1} we denote the last index of the set, e. g., for A_{011}, we have $B_{-1} = 1$. A decision stump is a 1-depth binary regression tree.

– Whereas traditionally the function space of decision stumps/trees were the *de facto* base classes, we can also use linear functions, splines, SVMs, shallow networks, etc.

Let us see an example of how to find an optimal $h \in \mathcal{H}_{DS}$ in the class of decision stumps.

Example 9.1.3. Let $\mathcal{X} = \mathbb{R}^d$ and consider the base hypothesis

$$\mathcal{H}_{DS} = \{x \mapsto \text{sign}(\theta - x_j)b : \theta \in \mathbb{R}, j \in \{1, \ldots, d\}, b \in \{\pm 1\}\}.$$

Note that the parameter b flips the signs of the labels; for simplicity, let $b = 1$. Let $S = \{(x_1, y_1), \ldots, (x_m, y_m)\} \subset \mathbb{R}^d \times \{-1, 1\}$ be a training set. We show how to implement an ERM rule, namely, how to find a decision stump that minimizes $L(h)$.

Each decision stump is parameterized by an index $j \in \{1, \ldots, d\}$ (selects which dimension of the feature vector to split over) and a threshold $\theta \in \mathbb{R}$. Thus minimizing $L(h)$ amounts to solving the minimization problem

$$\min_{j \in \{1, \ldots, d\}} \min_{\theta \in \mathbb{R}} \left(\sum_{i: y_i = 1} \mathbb{1}_{\{x_{i,j} \geq \theta\}} + \sum_{i: y_i = -1} \mathbb{1}_{\{x_{i,j} < \theta\}} \right), \tag{9.1}$$

where $x_{i,j}$ denotes the jth component of the ith data point. Here the empirical loss is expressed to show the mislabeling when the true label is positive and the prediction is negative, and when the true label is negative and the prediction is positive. We can simplify the above further.

Fix $j \in \{1, \ldots, d\}$ and relabel the training examples so that $x_{1,j} \leq x_{2,j} \leq \cdots \leq x_{m,j}$. Let us define the set $\Theta_j = \{\frac{x_{i,j} + x_{i+1,j}}{2} : i \in \{1, \ldots, m-1\}\} \cup \{x_{1,j} - 1, x_{m,j} + 1\}$. This sets up a grid for which, for any $\theta \in \mathbb{R}$, there exists $\theta' \in \Theta_j$ that yields the same predictions for the sample S. Then we can minimize θ over the set Θ_j.

This gives us an efficient procedure: choose $j \in \{1, \ldots, d\}$ and $\theta \in \Theta_j$ that minimize the objective value in (9.1).

9.2 AdaBoost

AdaBoost (short for Adaptive Boosting), informally, is an algorithm that combines classifiers from simple base classes to form a well-performing classifier.

9.2.1 Definition and performance guarantees

Definition 9.2.1 (Weighted empirical loss). Let $S = \{(x_i, y_i)\}_{i=1}^{m}$ be a dataset, and let $D \in \mathbb{R}^m$ be a probability vector, that is, all coordinates of D are nonnegative, and $\sum_{i \leq m} D_i = 1$. A weighted empirical loss for a binary classifier h is defined as

$$L_D(h) := \sum_{i=1}^{m} D_i \mathbb{1}_{\{h(x_i) \neq y_i\}}.$$

Note that if $D = (1/m, \dots, 1/m)$, then $L_D(h) = L(h)$.

By introducing a probability vector D we assign a weight to each individual prediction that weighs its contribution to the empirical error. Then we are able to express importance of getting one prediction right over another. This is a fundamental concept for the AdaBoost algorithm, which we now define.

Definition 9.2.2 (AdaBoost algorithm for binary classification). The AdaBoost algorithm takes as input a training set of examples $S = \{(x_1, y_1), \dots, (x_m, y_m)\}$, a base class $\mathcal{B}$, and $\gamma \in (0, 1/2]$. A classifier is built sequentially. Let the initial datapoint probability vector $D^{(1)} \in \mathbb{R}^m$ be given by $(1/m, \dots, 1/m)$. At each timestep $t = 1, \dots, T$ (also called round), the algorithm proceeds as follows:

1. Using the probability vector $D^{(t)}$ and the sample S, find $h_t \in \mathcal{B}$ such that $L_{D^{(t)}}(h_t) \leq 1/2 - \gamma$.
2. Compute the error of h_t, $\epsilon_t = L_{D^{(t)}}(h_t)$ and the classifier weight $w_t = \frac{1}{2} \log(\frac{1}{\epsilon_t} - 1)$.
3. Update the probability vector over the samples S, denoted by $D^{(t+1)} \in \mathbb{R}^m$,

$$D_i^{(t+1)} = \frac{D_i^{(t)} \exp(-w_t y_i h_t(x_i))}{\sum_{j=1}^{m} D_j^{(t)} \exp(-w_t y_j h_t(x_j))}, \quad i = 1, \dots, m.$$

After T iterations, the classifier returned by AdaBoost is

$$f_w(x) = \text{sign}\left(\sum_{t=1}^{T} w_t h_t(x)\right).$$

Note that if at any t, we have $\epsilon_t = 0$, then the weak learner h_t actually perfectly fits all datapoints. Since the principle of AdaBoost is to aggregate weak learners, we implicitly assume that this does not happen and $\epsilon_t > 0$ for all $t = 1, \dots, T$.

Remark 9.2.3. Although $D^{(t+1)}$ is computed using the performance of h_t and the probability vector $D^{(t)}$, it contains the information of the ensemble up until $t - 1$ through $D^{(t)}$. Indeed, $D_i^{(t+1)}$ is proportional to $\frac{1}{m} \exp(-y_i \sum_{k=1}^{t} w_k h_k(x_i))$. Thus the updated probability vector $D^{(t+1)}$ will give a higher probability mass to the training examples on which the previous classifiers $h_0, \dots, h_t$ were incorrect and a lower probability mass to training examples on which $h_1, \ldots, h_t$ were correct, thus, for example, giving the highest weight to the points that were misclassified by all previous classifiers.[1]

Remark 9.2.4. Besides $D^{(t)}$, which distributes masses at data points at each timestep t, another quantity plays a role of *weight*, namely, w_t. It gives the weights of the contribution of the base classifier h_t at timestep t in the final model f_w.

Remark 9.2.5. The number of timesteps/rounds T is often a hyperparameter of the AdaBoost algorithm.

What can be said of a classifier obtained using AdaBoost? In the following theorem, we show that the training error of a classifier generated using AdaBoost is upper bounded and exponentially decaying.

Theorem 9.2.6. *Let S be a training set and assume that at each iteration of AdaBoost, we obtain a hypothesis h_t for which $e_t \le 1/2 - \gamma$. Then the training error of the output hypothesis of AdaBoost f_w is at most*

$$L(f_w) = \frac{1}{m} \sum_{i=1}^{m} \mathbb{1}_{\{f_w(x_i) \neq y_i\}} \le \exp(-2\gamma^2 T).$$

Proof. For each round t, let $\tilde{f}_t = \sum_{k \le t} w_k h_k$, so that the output of AdaBoost is $f_w := \mathrm{sign}(\tilde{f}_T)$. In addition, let

$$Z_t := \sum_{i=1}^{m} D_i^{(t)} e^{-y_i w_t h_t(x_i)},$$

which is the normalization factor, so that $D_i^{(t+1)}$ is indeed a probability distribution.
Unrolling the recurrence to get $D^{(t+1)}$ from $D^{(t)}$, for all $i = 1, \ldots, m$, we can write

$$D_i^{(t+1)} = D_i^{(1)} \prod_{k=1}^{t} \frac{\exp(-w_k y_i h_k(x_i))}{Z_k}$$

$$= \frac{1}{m} \times \frac{\exp(-y_i \sum_{k=1}^{t} w_k h_k(x_i))}{\prod_{t=1}^{T} Z_t}$$

$$= \frac{1}{m} \times \frac{\exp(-y_i \tilde{f}_t(x_i))}{\prod_{k=1}^{t} Z_k}.$$

Note that $\mathbb{1}_{\{f_w(x) \neq y\}} \le e^{-y f_w(x)}$, since x is misclassified by f_w if and only if $\tilde{f}_T(x)$ and y have opposite signs. Therefore

$$L(f_w) = \frac{1}{m} \sum_{i=1}^{m} \mathbb{1}_{\{f_w(x_i) \neq y_i\}}$$

---

1 A video that shows the Adaboost algorithm: https://www.youtube.com/watch?v=k4G2VCuOMMg

$$\le \frac{1}{m} \sum_{i=1}^{m} \exp(-y_i \tilde{f}_T(x_i))$$

$$= \sum_{i=1}^{m} D_i^{(T+1)} \prod_{t=1}^{T} Z_t$$

$$= \prod_{t=1}^{T} Z_t.$$

We now rewrite Z_t as

$$Z_t = \sum_{i=1}^{m} D_i^{(t)} \exp(-w_t y_i h_t(x_i))$$

$$= \sum_{i:y_i=h_t(x_i)} D_i^{(t)} \exp(-w_t) + \sum_{i:y_i \ne h_t(x_i)} D_i^{(t)} \exp(w_t)$$

$$= (1 - \epsilon_t) \exp(-w_t) + \epsilon_t \exp(w_t)$$

by the definition of ϵ_t. Furthermore, by the definition of $w_t = \frac{1}{2} \log(1/\epsilon_t - 1)$, we get

$$Z_t = (1 - \epsilon_t) \sqrt{\frac{\epsilon_t}{1 - \epsilon_t}} + \epsilon_t \sqrt{\frac{1 - \epsilon_t}{\epsilon_t}}$$

$$= \sqrt{4\epsilon_t(1 - \epsilon_t)}.$$

By our assumption we have that $\epsilon_t \le 1/2 - \gamma$, and we can bound the quantity above:

$$Z_t \le \sqrt{1 - 4\gamma^2},$$

since $g(x) = x(1 - x)$ is increasing in $[0, 1/2]$. We thus have proven that

$$L(f_w) \le (1 - 4\gamma^2)^{T/2}.$$

To conclude, recall that for all $x \in \mathbb{R}$, the exponential function satisfies $1 + x \le \exp(x)$, so that $1 - 4\gamma^2 \le \exp(-4\gamma^2)$, which entails that

$$L(f_w) \le \exp(-2\gamma^2 T),$$

as claimed.   □

Thanks to Theorem 9.2.6, we see that even though the base classifier h_t performs only slightly better than a purely (uniform) random guess, AdaBoost is able to choose a linear combination of them that yields a predictor whose error on the dataset decreases exponentially fast in the number of base learners. However, our primary concern is the true risk of the output hypothesis, i. e., the generalization error. For example, if we drive the training error to zero (i. e., the number of rounds T grows), then will we overfit?

Although initially it was believed that yes [17], it has been observed that the test error can still decrease even when the training error is 0.

Because of this, a new analysis of the AdaBoost algorithm was given based on the concept of *margin*.

Definition 9.2.7 (Margin). For any datapoint (x, y), the margin for a classifier $f = \sum_t w_t h_t$ is defined as

$$m_f(x, y) = \frac{y \sum_t w_t h_t(x)}{\sum_t |w_t|}.$$

The margin is a number computed for any datapoint (x, y) that lies between -1 and 1. It is positive if and only if f classifies the example correctly, attaining 1 when all the base classifiers are correct and -1 when all the base classifiers misclassify the datapoint. Furthermore, the magnitude of the margin can be interpreted as a measure of confidence on the prediction, namely, if it is close to 0, then the weighted average of the predictions is split between classifying the point correctly and incorrectly. It was proved in [3] that larger margins on the training set translate into a better upper bound on the generalization error. In particular, the generalization error for any $\theta > 0$ is bounded with high probability:

$$R_D(f) \leq \mathbb{P}_{(x,y) \sim S}[m_f(x, y) \leq \theta] + O\left(\sqrt{\frac{d}{m\theta^2}}\right),$$

where d denotes the Vapnik–Chervonenkis (VC) dimension of the base hypothesis class (VC dimension is mentioned in Remark 4.2.5), m is the sample size of dataset S, and datapoint (x, y) is uniformly sampled from the training dataset S. Following the notion of margins, this bound has been explored and refined further [40].

9.2.2 Connections of AdaBoost to other models

We saw, through AdaBoost, that boosting essentially applies the weak classification algorithm to repeatedly modified versions of the data, producing a sequence of weak classifiers $h_t(x)$, $t = 1, \ldots, T$. The predictions are then combined through a weighted majority vote to produce the final prediction:

$$f_w(x) = \text{sign}\left(\sum_{t=1}^{T} w_t h_t(x)\right),$$

where w_t is computed by the boosting algorithm, weighting the contribution of each respective h_t. Their effect is to give higher influence to the more accurate classifiers in the sequence.

In this section, we will first show that AdaBoost belongs to the family of *additive models*, optimizing a specific loss function, the exponential loss function, following [19]. This equivalence was only understood a few years after AdaBoost inception.

Definition 9.2.8 (Additive models). An additive model f is a function of the form

$$f(x) = \sum_{t=1}^{T} w_t h(x; \psi_t),$$

where w_t are the expansion coefficients, and $h(x; \psi) \in \mathbb{R}$ are usually simple functions of the multivariate argument x, characterized by a set of parameters ψ.

An example of additive models is when each map $h(x; \psi)$ is a decision stump, and ψ parameterizes the split variables and split points. However, more general basis functions can be used to in additive models to generate a probability estimate for the class membership or regression models.

Additive models can be fit by minimizing a loss function ℓ averaged over the training data:

$$\min_{\{w_t, \psi_t\}_{t=1}^{T}} \sum_{i=1}^{m} \ell\left(y_i, \sum_{t=1}^{T} w_t h(x_i; \psi_t) \right). \tag{9.2}$$

However, the minimization problem (9.2) requires computationally intensive numerical optimization techniques. A simple alternative is to solve the subproblem of fitting just a single basis function at a time. This is called a *forward stagewise procedure* and aims to approximate the solution of (9.2) by sequentially adding new basis functions to the expansion without adjusting parameters and coefficients (w, ψ) of those that have been already added. This is outlined in Algorithm 9.1.

Algorithm 9.1: Forward stagewise additive modeling.

---

1 initialize $f_0(x) = 0$
2 **for** $t = 1, \ldots, T$ **do**
3     $\quad$ Compute $(w_t, \psi_t) = \arg\min_{w, \psi} \sum_{i=1}^{m} \ell(y_i, f_{k-1}(x_i) + wh(x_i; \psi))$
4     $\quad$ Set $f_t(x) = f_{t-1}(x) + w_t h(x; \psi_t)$
5 **end**

At each iteration t, we solve for the optimal basis function $h(x; \psi_t)$ and corresponding w_t to add to the current expansion $f_{t-1}(x)$. This produces $f_t(x)$, and the process is repeated. Previously added terms are not modified. Boosting can be seen as a way of fitting an additive expansion in a set of elementary "basis" functions, as shown in the next lemma.

Lemma 9.2.9. The AdaBoost algorithm (Definition 9.2.2) is equivalent to an additive model f solved via a forward stagewise procedure using the loss function

$$\ell(y, f(x)) = \exp(-yf(x))$$

and basis functions that map to $\{-1, 1\}$.

Proof. Considering the exponential loss, the tth step minimization reads

$$(w_t, \psi_t) = \arg\min_{w, \psi} \sum_{i=1}^{m} \exp(y_i(f_{t-1}(x_i) + wh(x_i; \psi))).$$

Let $\eta_i^{(t)} = \exp(-y_i(f_{t-1}(x_i)))$ for all $i \in \{1, \dots, m\}$. Then the loss can be written as

$$\sum_{i}^{m} \ell(y, f_t(x)) = \exp(-w) \sum_{i: y_i = h(x_i; \psi_t)} \eta_i^{(t)} + \exp(w) \sum_{i: y_i \neq h(x_i; \psi_t)} \eta_i^{(t)}$$

$$= \exp(-w) \sum_{i=1}^{m} \eta_i^{(t)} + (\exp(w) - \exp(-w)) \sum_{i=1}^{m} \eta_i^{(t)} \mathbb{1}_{\{y_i \neq h(x_i; \psi_t)\}}.$$

Solving the optimization sequentially, for a fixed weight w, solve for the parameter ψ

$$\hat{\psi} = \arg\min_{\psi} \sum_{i=1}^{m} \eta_i^{(t)} \mathbb{1}_{\{y_i \neq h(x_i; \psi)\}},$$

which recovers the predictor $h(x; \hat{\psi})$ that minimizes a prediction error weighted by $\eta_i^{(t)}$. Then, for given h, we can compute the derivative of the objective with respect to w, and setting it to 0 yields

$$0 = -\exp(-w) \sum_{i=1}^{m} \eta_i^{(t)} + (\exp(w) + \exp(-w)) \sum_{i=1}^{m} \eta_i^{(t)} \mathbb{1}_{\{y_i \neq h(x_i)\}}.$$

Multiplying the equation by $\exp(w)$ and using the fact that $\sum_{i=1}^{m} \eta_i^{(t)} = \sum_{i=1}^{m} \eta_i^{(t)} \mathbb{1}_{\{y_i = h(x_i)\}} + \sum_{i=1}^{m} \eta_i^{(t)} \mathbb{1}_{\{y_i \neq h(x_i)\}}$ yield

$$0 = \exp(2w) \sum_{i=1}^{m} \eta_i^{(t)} \mathbb{1}_{\{y_i \neq h(x_i)\}} - \sum_{i=1}^{m} \eta_i^{(t)} \mathbb{1}_{\{y_i = h(x_i)\}}.$$

Then we have

$$w = \frac{1}{2} \log\left(\frac{\sum_{i=1}^{m} \eta_i^{(t)} \mathbb{1}_{\{y_i = h(x_i)\}}}{\sum_{i=1}^{m} \eta_i^{(t)} \mathbb{1}_{\{y_i \neq h(x_i)\}}} \right).$$

Then we can recover the AdaBoost error by noting that

$$\frac{\eta_i^{(t)}}{\sum_{j=1}^m \eta_j^{(t)}} = D_i^{(t)}.$$

Then $\epsilon_t := \frac{\sum_{i=1}^m \eta_i^{(t)} \mathbb{1}_{\{y_i \neq h(x_i)\}}}{\sum_{j=1}^m \eta_j^{(t)}} = \sum_{i=1}^m D_i^{(t)} \mathbb{1}_{\{y_i \neq h(x_i)\}}$. Thus w becomes the familiar coefficient

$$w = \frac{1}{2} \log\left(\frac{1 - \epsilon_t}{\epsilon_t}\right).$$

At each stage of the forward stagewise procedure, another term that corresponds to the new weak classifier at each stage of the AdaBoost is added. □

9.2.3 AdaBoost for regression

Although in this chapter, we have focused mainly on binary classification tasks, solving multiclass and regression tasks with boosting is also possible. One of the first extensions of AdaBoost to regression tasks, called AdaBoost.R, is done in [18], where a regression problem is reduced to a binary classification problem, and then the familiar AdaBoost algorithm is applied, thus enjoying similar theoretical properties as the original AdaBoost algorithm. A subsequent modification of AdaBoost.R is presented in [16], which we discuss in this brief section as it is a popular method used in practice as the method implemented in the scikit-learn library.

Definition 9.2.10 (Weighted general empirical loss). Let $S = \{(x_i, y_i)\}_{i=1}^m$ be a dataset, and let $D \in \mathbb{R}^m$ be a probability vector, that is, all elements of D are nonnegative, and $\sum_{i \leq m} D_i = 1$. A weighted empirical loss for a general learner h is defined as

$$L_D(h; \ell) := \sum_{i=1}^m D_i \ell(y_i, h(x_i))$$

for any $\ell : \mathcal{Y} \times \mathcal{Y} \to R^+$.

Definition 9.2.11 (AdaBoost.R2). The AdaBoost.R2 algorithm takes as input a training set of examples $S = \{(x_1, y_1), \ldots, (x_m, y_m)\}$ and a base class $\mathcal{B}$. A regressor is built sequentially. Let the initial datapoint probability vector $D^{(0)} \in \mathbb{R}^m$ be given by $(1/m, \ldots, 1/m)$. At each timestep t, while the empirical loss $L_{D^{(t)}}(h_t; \ell)$ is smaller than 0.5 for a suitable loss ℓ with range in $[0, 1]$, the algorithm proceeds as follows:

1. Pick m samples from the dataset S to form the training set S_t by sampling with replacement, considering the sampling probability of each point to be given by $D_i^{(t)}$.
2. Train a predictor $h_t \in \mathcal{B}$ using the training set S_t.
3. Compute the loss for each training sample in S, denoting $\ell_i = \ell(y_i, h_t(x_i))$. Then compute the empirical loss $L_{D^{(t)}}(h_t; \ell)$.

4. Compute the *predictor weight* $w_t = \frac{L_{D^{(t)}}(h;\ell)}{1-L_{D^{(t)}}(h;\ell)}$.
5. Update the datapoint probability vector:

$$D_i^{(t+1)} = \frac{D_i^{(t)} w_t^{1-\ell_i}}{\sum_{j=1}^m D_j^{(t)} w_t^{1-\ell_j}}, \quad i = 1,\ldots,m.$$

At the end of iterations, for any input x, each of the T predictors h_t makes a prediction $h_t(x)$, yielding a set of ranked predictions $Y = \{h_{(1)}(x),\ldots,h_{(T)}(x)\}$, where $h_{(1)}(x_i) \leq h_{(2)}(x_i) \leq \cdots \leq h_{(T)}(x_i)$ (as well as the associated w_t). The approximator built by AdaBoost.R2 returns the value

$$f_w(x) = \inf\left\{y \in Y : \sum_{t:h_{(t)}\leq y} \log(1/w_{(t)}) \geq \frac{1}{2}\sum_t \log(1/w_{(t)})\right\}. \tag{9.3}$$

The response returned by the ensemble prediction by AdaBoost.R2 is a weighted median value.

To clarify some steps of the algorithm, let us make a few comments. Steps 1 and 2 could be replaced with training h_t on $L_{D^{(t)}}(h_t,\ell)$ directly instead of creating a new dataset S_t; however, the version above is the widely used algorithm introduced in [16]. In step 4, note that by definition we have $w_t < 1$. This weight w_t can be interpreted as a measure of confidence in the predictor; namely, in this case, a small w_t means a high confidence in the prediction by observing the effect of w_t in equation (9.3). Furthermore, as $w_t < 1$, we can see that in step 5, the smaller the loss ℓ_i, the smaller the probability $D_i^{(t+1)}$, that is, the probability that point i will be picked as a member of the training set S_1 in the next iteration.

Remark 9.2.12. In the definition of AdaBoost.R2, we require the loss range to be in $[0,1]$. A common way to guarantee this is computing the quantity $N := \sup_{(x_i,y_i)\in S_1} |h_t(x_i) - y_i|$. Then common loss choices are given as
-   $\ell_i = |h_t(x_i) - y_i|/N$, which yield a linear loss,
-   $\ell_i = |h_t(x_i) - y_i|^2/N^2$, a square loss, and
-   $\ell_i = 1 - \exp(-|h_t(x_i) - y_i|/N)$, an exponential loss.

As mentioned, although this is a popular method in practice, by definition we have no guarantee that the training error goes to zero as the number of rounds increases. In particular, as the datapoint probability for some point i, $D_i^{(t)}$ increases, which means that *difficult* examples are more likely to be chosen for the next iteration training set. Each subsequent approximator has a harder set of examples to train on. Thus the empirical loss of the last weak learner tends to increase as the algorithm continues, and ultimately, when the bound $L_{D^{(t)}}(h_t;\ell) < 0.5$ is not satisfied, it leads to the termination of the algorithm. Furthermore, there is no guarantee that the algorithm will always terminate.

An example of using AdaBoost.R2 for regression to learn a quadratic function can be found in Figure 9.2

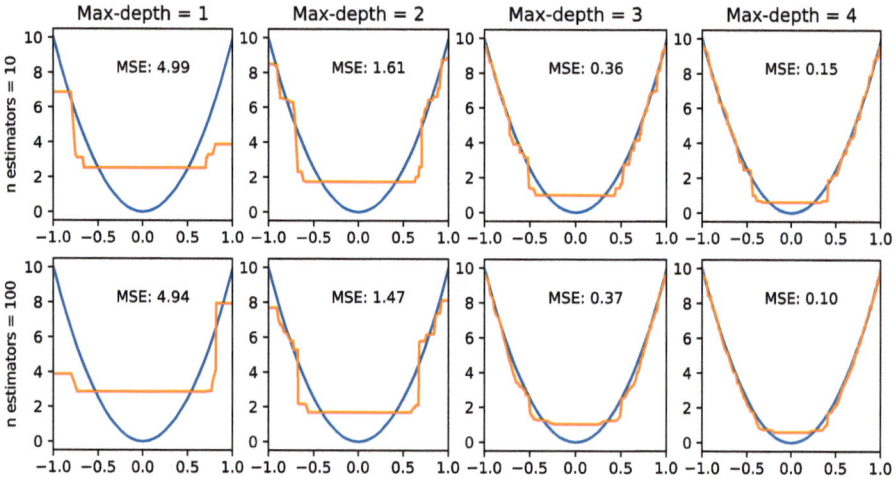

Figure 9.2: Example of AdaBoost for regression tasks. Max-depth = 1 corresponds to decision stumps as the base learner.

9.3 Gradient boosting

We finish this chapter by briefly introducing Gradient Boosting, an algorithm introduced in [20] as a greedy function approximation after AdaBoost. Similarly to AdaBoost, Gradient Boosting can be thought of as an additive model of the form

$$f_T(x) = \sum_{i=0}^{T} w_t h_t(x).$$

The main difference between AdaBoost and Gradient Boosting is that the predictor added at time t is not introduced by finding a good predictor on the reweighted data, but rather through approximating *gradient steps*.

Namely, let us consider the following quantity at some discrete time t:

$$g_t(x_i) = -\left[\frac{\partial \ell(y_i, f(x_i))}{\partial f(x_i)}\right]_{f(x)=f_{t-1}(x)}, \quad i = 1, \ldots, m.$$

This gives an approximation of the best steepest-descent step direction by considering m gradient estimates $\vec{g}_t = \{g_t(x_i)\}_i^m$ at $f_{t-1}(x)$. The idea of gradient boosting is to approximate this quantity, thus mimicking the gradient descent step. One first clear advantage of this formulation is that we can consider a general loss function ℓ as long as it is differentiable. Another advantage is that it is computationally efficient. Let us now define the algorithm.

Definition 9.3.1 (Gradient Boosting). The Gradient Boosting algorithm takes as input a training set of examples $S = \{(x_1, y_1), \ldots, (x_m, y_m)\}$ and a base class $\mathcal{B}$. A regression is built sequentially. At initialization we have

- $w_0 = \arg\min_\rho \sum_{i=1}^m \ell(y_i, \rho)$,
- $h_0(x) \equiv 1$,

yielding the constant initial approximation $f_0(x) = w_0$. Then for $t = 1, \ldots, T$, we have:

1. Compute the steepest-descent step direction for each training sample:

$$\tilde{y}_i = -\left[\frac{\partial \ell(y_i, f(x_i))}{\partial f(x_i)}\right]_{f(x)=f_{t-1}(x)}, \quad i = 1, \ldots, m.$$

2. Find a function h_t that best approximates the steepest-descent step direction:

$$h_t, \beta_t = \arg\min_{h,\beta} \sum_{i=1}^m (\tilde{y}_i - \beta h(x_i)).$$

3. Compute the coefficient that minimizes the loss (this can be interpreted as the stepsize in the gradient descent):

$$w_t = \arg\min_w \sum_{i=1}^m \ell(y_i, f_{t-1}(x_i) + w h_t(x_i)).$$

4. Update the model: $f_t(x) = f_{t-1}(x) + w_t h_t(x)$.

At the end of the iterations, the method returns the predictor

$$f_T(x) = \sum_{i=0}^T w_t h_t(x).$$

Remark 9.3.2. If the loss function is the square-loss function $\ell(x, y) = \frac{1}{2}\|x - y\|^2$, then the quantities $\tilde{y}_i$ can be interpreted as the differences between the predictor f_{t-1} and the true label y_i. Then step 3 can be skipped as the coefficient β_t is equivalent to w_t.

With Gradient Boosting, the task of classification or regression boils down to the appropriate choices of loss ℓ and base classifier $\mathcal{B}$ with a possible final transformation on f_T to yield, for instance, a label or a probability estimate (see details in [20]).

9.4 Implementation details

In this section, we present some code snippets that use AdaBoost and ensemble methods to solve problems in classification and regression.

Listing 9.1: Linear regression model using scikit-learn

```
 1 from sklearn.model_selection import train_test_split
 2 from sklearn.datasets import load_iris
 3 from sklearn.ensemble import AdaBoostClassifier
 4
 5 X, y = load_iris(return_X_y=True)
 6 X_train, X_test, y_train, y_test = train_test_split(X, y,
       random_state=0)
 7
 8 classifier = AdaBoostClassifier(n_estimators=100)
 9 classifier.fit(X_train, y_train)
10 y_pred = classifier.predict(X_test)
11
12 print(f"Mean accuracy: {classifier.score(X_test,y_test)}")
```

Listing 9.2: Linear regression model using scikit-learn

```
 1 import numpy as np
 2 from sklearn.ensemble import AdaBoostRegressor
 3 from sklearn.tree import DecisionTreeRegressor
 4
 5 # Generating data, learning the quadratic function in
       [-1,1]
 6 X_train = np.linspace(-1, 1, 100)[:,np.newaxis]
 7 y_train = X.ravel()**2
 8 X_test = np.array(sorted(np.random.uniform(low=-1,high=1,
       size=200)))[:,np.newaxis]
 9
10 regressor = AdaBoostRegressor(DecisionTreeRegressor(
       max_depth=3),n_estimators=30)
11                 # Taking DecisionTrees with maximum depth 3 to
       be the base learners
12
13 regressor.fit(X_train, y_train)
14 print(f"r2-score: {regressor.score(X_test,X_test.ravel()
       **2)}")
```

For more details and examples, visit the code repository: https://github.com/hanveiga/tmml

Part III: **Beyond supervised learning**

10 Topics in unsupervised learning

So far, we have always assumed that we have access to a data-generating process D that provides a pair of inputs and outputs (x, y), thus providing labeled data. In this section, we assume that we no longer have access to labeled data. We have access to a set of datapoints $\mathcal{X} = \{x_i\}_{i=1}^{m}$ and want to make sense of them. More formally, unsupervised learning tasks often correspond to minimizing a loss function of the form

$$L(f) = \frac{1}{m} \sum_{i=1}^{m} \ell(f(x_i), x_i),$$

where $\{x_i; i = 1, \ldots, m\}$ is the unlabeled dataset, and ℓ is a nonnegative map measuring the error at one datapoint. This is similar to the empirical risk minimization framework in supervised learning, that is, (1.3) with x_i in place of the label y_i. In this chapter, we discuss two families of unsupervised learning tasks, *clustering* and *dimensionality reduction*. Both of them fall into the above formalism.

Clustering is used often for exploratory data analysis, because it can find similarities and groupings between unlabeled datapoints. For example, suppose you get a set of articles from newspapers and you want to organize them by topic: you want to find natural groupings based, for instance, on the dominant topic.

Dimensionality reduction is used when the elements of the dataset lives in a very high-dimensional space $\mathcal{X}$, but the data actually lie on a lower-dimensional manifold embedded in the high-dimensional space. The dimensionality reduction task consists in finding such a low-dimensional space. This is the case where the data are composed of many possibly highly correlated variables, sometimes more numerous than the number of datapoints. For example, in DNA sequencing, several thousands of genes can be sequenced for one observation.

10.1 Clustering

We can think of the task of clustering as grouping sets of objects such that similar objects are in the same group (cluster) and dissimilar objects are in different clusters. There are, however, some difficulties on how to formally define similarity and dissimilarity. For instance, in Figure 10.1, two different notions of similarity will lead to two distinct groupings. In the first grouping, similarity means that points that are close to each other are similar and belong to the same cluster, whereas in the second grouping, we find clusters that minimize the interdistance between the points inside the same cluster.

These are two important challenges with clustering:

- **Ambiguity in the notion of similarity:** The two aforementioned views seem to contradict each other.

https://doi.org/10.1515/9783111288994-010

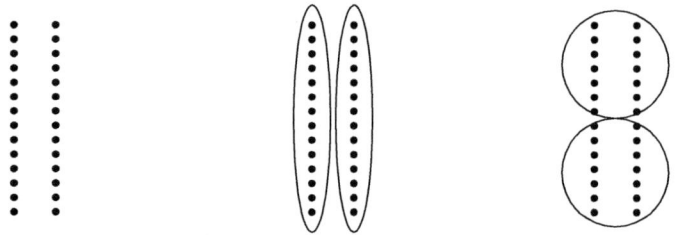

Figure 10.1: Cluster assignment under different notions of similarity.

– **Lack of ground truth:** Often, there is no true label to the cluster membership, and thus the evaluation of the clustering strategy can be "subjective" and cumbersome.

10.1.1 Definitions

Let us formalize a clustering model with the following definition. In this section, we consider a metric space $\mathcal{X}$ endowed with a metric $d : \mathcal{X} \times \mathcal{X} \to \mathbb{R}_0^+$. The set of subsets of $\mathcal{X}$ is denoted by $S(\mathcal{X}) = \{B \subset \mathcal{X}\}$.

Definition 10.1.1 (Clustering model). We say that a function $f : \mathcal{X} \to S(\mathcal{X})$ is a *clustering model* on $\mathcal{X}$ if it satisfies
– $x \in f(x) \ \forall x \in \mathcal{X}$.
– For all $x, x' \in \mathcal{X}$, either $f(x) = f(x')$, or $f(x) \cap f(x') = \emptyset$.

The set of clusters of f is denoted by $C_f := \{f(x); x \in \mathcal{X}\}$ (without repetition). It is a partition of $\mathcal{X}$.

We also make clear that $f(x_i)$ returns all points in the cluster for which x_i belongs to. For example, let us consider the set $\mathcal{X} = \{x_1, x_2, x_3, x_4, x_5\}$ and some clustering function f that returns two clusters $C_f = \{\{x_1, x_2, x_3\}, \{x_4, x_5\}\}$. Then, for example, the output of the clustering function at x_1 is $f(x_1) = \{x_1, x_2, x_3\}$.

Remark 10.1.2. The number of clusters $|C_f|$ is often a quantity that has to be specified.

In the supervised learning formalism, we considered a data distribution D generating i. i. d. inputs and labels (x_i, y_i). The empirical risk minimization framework used the dataset $\{(x_i, y_i)\}_{i=1}^{m}$ as an approximation of the real data distribution, so that we could use the empirical loss as a proxy of the *true* loss. It is possible to reason similarly in the unsupervised learning framework, but we refrain to do so and instead work directly with a finite set of inputs for the following reasons: a) We are generally given a finite set of objects, which is not seen as a proxy to a larger set, and b) the inputs are often not independent. Therefore we write the objective to optimize a clustering task *directly* on a finite set.

We consider a finite set $x = \{x_i\}_{i=1}^m \subset \mathcal{X}$. The set of partitions of x is denoted by $\mathcal{P}(x)$.

Definition 10.1.3. Let $C(x)$ be the set of clustering models on x, and let $\tilde{C}(x) \subset C(x)$ be a subset of clustering models. A clustering task corresponds to solving

$$\arg\min_{f \in \tilde{C}(x)} L(C_f, x)$$

for a well-chosen loss function $L : \mathcal{P}(x) \times x \to \mathbb{R}_+$.

For now, the above definition is abstract since the dependence of L on the metric d is implicit, as well as the choice of the valid clusterings $\tilde{C}(x)$. Note that we can then perform a classification to generalize outside x using a clustering model f. Namely, for a new point $x \in \mathcal{X}$ that is not in x, we can attribute it to the right cluster membership. One way is solving a supervised learning task on the now <u>labeled</u> dataset $\{(x_i, g(f(x_i))); i = 1, \dots, m\}$, where $g : \mathcal{P}(x) \to \mathbb{N}$ returns an identifier to any cluster in C_f. Another way is to iteratively attribute new points to the closest present clusters. In this book, we thus prefer to consider the clustering and the classification tasks as two separate endeavors.

10.1.2 Procedural clustering

A clustering algorithm assumes that points that are close to each other according to the metric d belong to the same cluster. The general procedural algorithm proceeds as follows. Given a set of points $x = \{x_i\}_{i=1}^m$,
1. Each point starts in its own cluster.
2. Iteratively, based on the distances between clusters constructed from d, two nearby clusters merge into one cluster.
3. Ends according to some stopping criterion.

Thus the number of clusters decreases at each iteration of the algorithm. In the limit, one large cluster is generated if there are no additional stopping criteria. This procedure is said to be *agglomerative*. Starting from one big cluster and splitting procedurally are also possible, and in that case the algorithm is said to be *divisive*.

There are several ways for the metric d to induce a distance between two clusters A and B. Common distances are
- Single linkage:

$$D(A, B) = \min\{d(x, y) : x \in A, y \in B\}.$$

- Average linkage:

$$D(A, B) = \frac{1}{|A||B|} \sum_{x \in A} \sum_{y \in B} d(x, y).$$

– Maximum linkage:

$$D(A,B) = \max\{d(x,y) : x \in A, y \in B\}.$$

As mentioned above, without stopping criterion, procedural clustering ends at the trivial clustering $f(x_i) = \{x_i\}_{i=1}^m$. Hence we need to constrain the clustering models. Common restrictions are, for instance, prescribing the number of clusters or setting a distance $\delta > 0$ such that the algorithm stops when all clusters are at distance at least δ from each other. (Mixtures of the two conditions are also used.) In what follows, we will only consider the set of clustering models on x with exactly k clusters, denoted by $\mathcal{C}_k(x)$.

For $B \subset x$, let $\ell(B) := 1/D(x \setminus B, B)$ with the convention that $D(\emptyset, B) = \infty$ and $1/\infty = 0$. Let $L(f) := \max_{B \in \mathcal{C}_f} \ell(B)$. Then the clustering can be written as

$$\arg \min_{f \in \mathcal{C}_k(x)} L(f). \tag{10.1}$$

By seeking the argmin of $L(f)$ we find a clustering function that maximizes the distance between the closest two clusters: consider $x = \{1, 5, 10\}$ and the Euclidean distance d on $\mathbb{R}$; then if f, g are such that $C_f = \{\{1, 5\}, \{10\}\}$ and $C_g = \{\{1\}, \{5, 10\}\}$, then we see that $L(f) = 1/d(5, 10) = 1/5$ and $L(g) = 1/d(1, 5) = 1/4$, so that L favors f over g.

Formally, the procedural agglomerative algorithm is as follows: if $k \le m$, then
– $f_0(x_i) = \{x_i\}$ for all $i = 1, \ldots, m$.
– For $n = 0, \ldots, m - k - 1$, take $B_1 \ne B_2 \in C_{f_n}$ that minimize $D(B_1, B_2)$ and merge B_1 and B_2, that is, $f_{n+1}(x_i) = B_1 \cup B_2$ if $x_i \in B_1 \cup B_2$ and $f_{n+1}(x_i) = f_n(x_i)$ otherwise.

The possible ambiguities (ties in minimal $D(B_1, B_2)$) can be lifted in any arbitrary way.

Theorem 10.1.4. *If D is the single linkage or the maximum linkage distance, then the hierarchical clustering algorithm converges to a solution of* (10.1). *Moreover, if the distances $\{d(x_i, x_j); 1 \le i < j \le m\}$ are pairwise distinct, then the optimum is unique.*

The average linkage distance is harder to study because the distance between two clusters depends on all points within that cluster. We refer the curious reader to Chapter 8 of [32] for more detail on clustering methods, motivated by ecology, with a general mathematical formalism.

Proof. We assume that $\{d(x_i, x_j); 1 \le i < j \le m\}$ are pairwise distinct, the case where it is not satisfied being easily treated by setting an arbitrary rule to get rid of possible ties. To ease the presentation, we only consider the single linkage distance $D(A, B) = \min\{d(x, y) : x \in A, y \in B\}$.

We claim that an optimal clustering f_*, solution of (10.1), must satisfy the following: the set of possible splittings of a cluster of f_* into two nonempty clusters is finite. Therefore there exists a splitting $A_1 \cup A_2 = f_*(x_{i_0})$ such that $D(A_1, A_2)$ is minimal among splittings. Such a splitting satisfies

$$D(A_1, A_2) < D(B_1, B_2) \quad \forall B_1 \neq B_2 \in C_{f_*}. \tag{10.2}$$

Indeed, suppose that it is not the case, that is, there exist $B_1 \neq B_2 \in C_{f_*}$ such that $L(f_*) = 1/D(B_1, B_2)$ and $D(B_1, B_2) < D(A_1, A_2)$ (the distances are pairwise distinct by assumption). Let f' be the clustering obtained from f by splitting A_1 and A_2. Note that $L(f') = 1/D(B_1, B_2) = L(f_*)$. Then construct f'' by merging B_1 and B_2. Since we assume that $\{d(x_i, x_j); 1 \leq i < j \leq m\}$ are pairwise distinct, necessarily, $L(f'') < L(f')$, which contradicts the fact that f_* is the optimum clustering with k clusters, since f'' also contains k clusters. Therefore f_* has a unique minimal splitting that satisfies (10.2).

Let f'_* be the cluster obtained after performing this minimal splitting. It is easy to see that the optimal merging of this clustering, that is, merging $B_1 \neq B_2 \in C_{f'_*}$ such that $D(B_1, B_2)$ is minimal, yields the clustering f_*. By induction we get that the sequence of minimal splitting starting from f_* until reaching $\{\{x_1\}, \ldots, \{x_m\}\}$ is reversible, and the reverse procedure sequentially performs optimal merging as described above. The sequential merging is precisely what the agglomerative clustering algorithm does, which shows the claim and concludes the proof. $\qquad \square$

Note that the same argument can be applied to divisive procedural algorithms, i. e., starting from a single cluster and splitting it optimally iteratively until obtaining k clusters.

Remark 10.1.5. The procedural clustering algorithm presented in this chapter outputs a clustering with given number of clusters of a finite set of points x. More generally, we can be interested in the whole sequence of clusterings obtained from the initial clustering $f_0 = \{\{x_1\}, \ldots, \{x_m\}\}$ to $f_m = \{\{x_1, \ldots, x_m\}\}$. This is, for instance, the case in Phylogeny, where we can be interested in relating different species based on genetic distances. A *hierarchical clustering algorithm* outputs a tree whose set of leaves is x and internal nodes are clusters. Using a procedural method, we can define a hierarchical clustering as the tree obtained by keeping track of all steps such that the nodes of the tree at height n are the clusters of the procedural algorithm at step n. Recent efforts were made to frame different hierarchical clustering algorithms as an optimization problem with a loss function on trees, which we believe were initiated by [14].

10.1.3 Centroid-based clustering

Recall that the set of points $x = \{x_i\}_{i=1}^m$ is embedded in a metric space $(\mathcal{X}, d)$. The *centroid of a cluster* C_i is defined as

$$\mu(C_i) = \arg\min_{\mu \in \mathcal{X}} \sum_{x \in C_i} d(x, \mu).$$

A common distance to consider is the squared distance $d(x, \mu) = \|x - \mu\|^2$. Note that $\mu(C_i)$ needs not be – and in general is not – an element of C_i.

Remark 10.1.6. Other distances can be defined, for example,

$$\mu(C_i) = \arg\min_{\mu \in \mathcal{X}} \sum_{x \in C_i} |x - \mu|,$$

which instead of considering the squared distance, considers the L_1 distance. These will, in turn, yield a different objective and solution.

In the previous section, the loss function was calculated from the distances between clusters. Given a notion of distance and the centroid definition, we define a loss function as follows: for a clustering, f and its clusters C_f, define

$$L(C_f, x) := \sum_{i=1}^{m} \ell(f(x_i), x_i),$$

where $\quad \ell(f(x_i), x_i) := d(\mu(f(x_i)), x_i).$

We seek to find the point membership to the different clusters $(C_1, \ldots, C_k)$ that minimizes the quantity above, namely, we seek to solve

$$\arg\min_{f \in \tilde{\mathcal{C}}} L(C_f, x), \tag{10.3}$$

where $\tilde{\mathcal{C}} \subset \mathcal{C}$ is a chosen set of valid clusterings, as before.

In this short section, we focus on the so-called *k-means clustering*. When points $x \subset \mathbb{R}^n$, it uses the squared distance $d(x, y) = \|x - y\|^2$. Using this distance, we can write the *k*-means objective function as follows:

$$L_{k\text{-means}}(C_f, x) = \sum_{i=1}^{k} \sum_{x \in C_i} \|\mu(C_i) - x\|^2,$$

where the number of clusters k is fixed, and we aim to solve the minimization problem as in (10.3).

Remark 10.1.7. Another notion of "center" of a cluster can be defined:

$$\mu(C_i) = \arg\min_{\mu \in \{x_j\}_{j=1}^{m}} \sum_{x \in C_i} \|x - \mu\|^2,$$

which is called the medoid. In this case, this quantity is a point in C_i.

In general, the minimization problem (10.3) is not easy to solve because the space of solutions is typically too large to search exhaustively, and since the centroid-based clustering presented above is non-parametric, gradient-based methods do not apply to optimize this particular loss. This also means that the *k-means objective* is not easy to solve. An algorithm, called the *k-means algorithm*, has been derived to approximate the solution to it. We describe it in Algorithm 10.1.

Algorithm 10.1: The k-means algorithm.

Input: A set of points $\{x_i\}_{i=1}^m \subset \mathcal{X} \subset \mathcal{R}^n$, number of clusters k

Output: $C = \{C_1, C_2, \ldots, C_k\}$ where each C_i is a set containing points x that belong to that cluster

1 initialize positions $\mu_1, \ldots, \mu_k$ randomly in $\mathbb{R}^n$

2 **while** *no convergence* **do**

3   $\quad \forall i \in [k] : C_i^{(t)} = \{x \in \{x_i\}_{i=1}^m : i = \arg\min_j \|x - \mu_j^{(t-1)}\|^2\}$

4   $\quad \forall i \in [k] : \mu_i^{(t)} = \frac{1}{|C_i|} \sum_{x \in C_i^{(t)}} x.$

5 **end**

This algorithms converges to a solution, but it neither solves the k-means objective nor converges to a unique cluster assignment due to the randomness in the initialization. However, we have the following lemma.

Proposition 10.1.8. *Each iteration of the k-means algorithm does not increase the k-means objective function.*

Proof. Let us consider the algorithm at time step t. Let $\mu_i^{(t-1)} = \mu(C_i^{(t-1)})$. The k-means objective, evaluated at the clusters from $t - 1$, reads

$$L_{k\text{-means}}(C_f^{(t-1)}, \mathcal{X}) = \sum_{i=1}^{k} \sum_{x \in C_i^{(t-1)}} \|\mu_i^{(t-1)} - x\|^2.$$

Following the algorithm, we have

$$\forall i \in [k] : C_i^{(t)} = \{x \in \{x_i\}_{i=1}^m : i = \arg\min_j \|x - \mu_j^{(t-1)}\|^2\},$$

yielding a new partition of $\mathcal{X}$, for which by definition

$$\sum_{i=1}^{k} \sum_{x \in C_i^{(t)}} \|\mu_i^{(t-1)} - x\|^2 \leq L_{k\text{-means}}(C_f^{(t-1)}, \mathcal{X}).$$

The new centroids are attained as

$$\mu_i^{(t)} = \frac{1}{|C_i^{(t)}|} \sum_{x \in C_i^{(t)}} x,$$

which is the solution to $\arg\min_\mu \sum_{x \in C_i^{(t)}} \|\mu - x\|^2$, and thus

$$\sum_{i=1}^{k} \sum_{x \in C_i^{(t)}} \|\mu_i^{(t-1)} - x\|^2 \geq \sum_{i=1}^{k} \sum_{x \in C_i^{(t)}} \|\mu_i^{(t)} - x\|^2 = L_{k\text{-means}}(C_f^{(t)}, \mathcal{X}).$$

$\square$

10.2 Dimensionality reduction

Dimensionality reduction can be motivated by the fact that data might lie in a low-dimensional manifold embedded in a high-dimensional space, as well as some practical concerns, such as the desire to improve computational performance, reduce data storage, and improve interpretability of the data.

Let us assume that we have a dataset $\mathcal{X} = \{x_i\}_{i=1}^m \subset \mathcal{X} \subset \mathbb{R}^n$. We want to find a representation of $\mathcal{X}$ in a lower-dimensional space $\mathbb{R}^r$, where $r \ll n$. Formally, we wish to find a *compression* function $f : \mathbb{R}^n \to \mathbb{R}^r$ such that, for some *reconstruction* $g : \mathbb{R}^r \to \mathbb{R}^n$, we have

$$g(f(x)) \approx x$$

for all $x \in \mathcal{X}$, meaning that the compression of x by f enables us a *good* recovery by g. In this section, we describe the *principal component analysis (PCA) method* and a kernelized version of it.

10.2.1 Principal component analysis

The principal component analysis (PCA) method considers functions f given by linear maps, i. e., f is a matrix $W \in \mathbb{R}^{r \times n}$, and g is a matrix $U \in \mathbb{R}^{n \times r}$. Then the PCA technique finds the matrices W and U that solve the minimization problem

$$\underset{\substack{W \in \mathbb{R}^{r \times n} \\ U \in \mathbb{R}^{n \times r}}}{\arg \min} \sum_{i=1}^m \|x_i - UWx_i\|^2, \tag{10.4}$$

which minimizes in a squared distance sense the reconstruction error of U given a compression by W. We may ask what can be said about matrices W and U, and how do we solve this optimization problem? It turns out that there are easy and elegant ways to compute these matrices W and U. The main result of this section is summarized in the following theorem, giving us an explicit form for the matrices W and U based on $\mathcal{X}$, which we will prove toward the end of this section.

Theorem 10.2.1. *Let $\mathcal{X} = \{x_i\}_{i=1}^m$ be a set of vectors in $\mathbb{R}^n$. Let $A = \sum_{i=1}^m x_i x_i^T \in \mathbb{R}^{n \times n}$ and $X = [x_1, \ldots, x_m] \in \mathbb{R}^{n \times m}$. Then*
1. *A is a symmetric positive semidefinite matrix.*
2. *Let $(v_1, \ldots, v_r)$ be the r normalized eigenvectors of A corresponding to the r largest eigenvalues ordered by decreasing size. Then the matrix $U = (v_1, \ldots, v_r)$ taking the eigenvectors of A as columns and $W = U^T$ solve the minimization problem (10.4).*
3. *Let the singular decomposition of X be given by $Q\Sigma P^T$. If U contains the first r columns of Q, and if $W = U^T$, then they solve the minimization problem (10.4).*

The theorem gives us two ways to compute the compression and reconstruction matrices W and U. We also note the solutions obtained via items 2 and 3 are the same. We first start by showing the relation between the compression matrix W and reconstruction matrix U in the following lemma.

Lemma 10.2.2. *Let (U, W) be a solution to* (10.4). *Then the columns of U are orthonormal, and $W = U^T$.*

Proof. Consider any W and U and the function $f(g(x)) = WUx$. The range R of $f(g(x))$ is an r-dimensional linear subspace of $\mathbb{R}^n$ with a basis of r vectors in $\mathbb{R}^n$. Let $V \in \mathbb{R}^{n \times r}$ be a matrix whose columns are an orthonormal basis of R. Any vector in R can be written as Va with $a \in \mathbb{R}^r$. Then, for all $x \in \mathbb{R}^n$ and $a \in \mathbb{R}^r$, we have

$$\|x - Va\|^2 = \|x\|^2 + \|a\|^2 - 2a^T V^T x,$$

as $V^T V$ is the identity in $\mathbb{R}^r$, which we henceforth denote by $\mathcal{I}_r$. Taking the gradient of the above with respect to a and setting it to zero yield

$$a = V^T x.$$

Therefore, for each $x \in \mathbb{R}^n$, we have

$$Va = VV^T x = \arg\min_{\tilde{x} \in R} \|x - \tilde{x}\|^2.$$

Because this holds for every $x_1, \ldots, x_m$, we can lower bound the objective

$$\sum_{i=1}^m \|x_i - U'W'x_i\|^2 \geq \sum_{i=1}^m \|x_i - VV^T x_i\|^2$$

for all U' and W', and hence the result follows. $\qquad\square$

Using the fact that $W = U^T$, we can rewrite the objective.

Lemma 10.2.3. *Objective* (10.4) *can be rewritten as*

$$\arg\max_{U \in \mathbb{R}^{n \times r} : U^T U = \mathcal{I}_r} \; tr\left(U^T \left(\sum_{i=1}^m x_i x_i^T \right) U \right),$$

where tr denotes the trace operator on matrices, i. e., $tr(M) = \sum_i M_{ii}$.

Proof. Using the previous lemma, we can write (10.4) as

$$\arg\min_{U \in \mathbb{R}^{n \times r} : U^T U = \mathcal{I}_r} \sum_{i=1}^m \|x_i - UU^T x_i\|^2.$$

Then, for all $x \in \mathbb{R}^n$ and orthonormal matrices $U \in \mathbb{R}^{n \times r}$, expanding the above we have

$$\left\| x_i - UU^T x_i \right\|^2 = \|x_i\|^2 - 2x_i^T UU^T x_i + x_i^T UU^T UU^T x_i$$
$$= \|x_i\|^2 - x_i^T UU^T x_i$$
$$= \|x_i\|^2 - tr(U^T x_i x_i^T U),$$

where we used a basic property of the trace for the last equality. Because the trace operator tr is a linear operator and we want to minimize the quantity above with respect to U, we can rewrite the objective as maximization

$$\underset{U \in \mathbb{R}^{n \times r}: U^T U = \mathcal{I}_r}{\arg \max} \; tr\left(U^T \left(\sum_{i=1}^m x_i x_i^T \right) U \right),$$

yielding the result. □

Let $A = \sum_{i=1}^m x_i x_i^T$ as in Theorem 10.2.1. We are now ready to prove the main theorem of this section, Theorem 10.2.1.

Proof.
1. We verify that A is semipositive definite: for any vector $a \in \mathbb{R}^n$, we have

$$a^T \left(\sum_{i=1}^m x_i x_i^T \right) a = \sum_{i=1}^m a^T (x_i x_i^T) a = \sum_{i=1}^m (a^T x_i)^T a^T x_i \geq 0.$$

2. Let $U \in \mathbb{R}^{n \times r}$ be an arbitrary matrix with orthonormal columns such that $U^T U = \mathcal{I}_r$. We write the spectral decomposition of $A = \sum_{i=1}^n \lambda_i v_i v_i^T$, yielding the objective:

$$tr(U^T A U) = \sum_{i=1}^r u_i^T \left(\sum_{j=1}^n \lambda_j v_j v_j^T \right) u_i$$
$$= \sum_{j=1}^n \lambda_j \sum_{i=1}^r (v_j^T u_i)^2$$
$$= \sum_{j=1}^n \lambda_j c_j, \qquad\qquad (10.5)$$

where $c_j := \sum_{i=1}^r (v_j^T u_i)^2$. Now we show that i) $c_j \in [0,1]$ for all $j \leq n$ and ii) $\sum_{j=1}^n c_j = r$.
i) The vectors $(u_1, \ldots, u_r)$ form an orthonormal basis of a linear subspace of $\mathbb{R}^n$ of dimension r. We complete the basis by adding vectors $u_{r+1}, \ldots, u_n$ such that $(u_1, \ldots, u_n)$ is an orthonormal basis of $\mathbb{R}^n$. Then any eigenvector v_j admits a representation with the basis above, that is, $v_j = \sum_{i=1}^n (v_j^T u_i) u_i$ for all $j \leq n$. Then

$$v_j^T v_j = \sum_{i=1}^n (v_j^T u_i)^2 = 1 \geq c_j = \sum_{i=1}^r (v_j^T u_i)^2 \geq 0,$$

first by the orthogonality of the u basis, and second by v_j being normalized.

ii) We write

$$\sum_{j=1}^{n} c_j = \sum_{j=1}^{n} \sum_{i=1}^{r} (v_j^T u_i)^2 = \sum_{i=1}^{r} \sum_{j=1}^{n} (v_j^T u_i)^2 = r$$

by using the same argument as above, since we can project any vector $u \in \mathbb{R}^n$ onto the basis $\{v_1, \ldots, v_n\}$, and the vectors u_i, $i \le n$, are normalized.

Then the maximum of (10.5) over all choices of $c_1, \ldots, c_n \in [0, 1]$ with $\sum_{j=1}^{n} c_j = r$ is $\lambda_1 + \lambda_2 + \cdots + \lambda_r$. This is achieved by $c_1 = c_2 = \cdots = c_r = 1$ and $c_{r+1} = \cdots = c_n = 0$, i. e., when span$\{u_1, \ldots, u_r\} =$ span$\{v_1, \ldots, v_r\}$, that is, we can choose $u_1, \ldots, u_r$ to be $v_1, \ldots, v_r$.

3.   Let $X = [x_1, \ldots, x_m]$ with a singular decomposition given by $Q\Sigma P^T$. Noting that $A = XX^T$, we have $A = Q\Sigma P^T (Q\Sigma P^T)^T$. Simplifying the expression

$$A = Q\Sigma P^T (Q\Sigma P^T)^T$$
$$= Q\Sigma P^T P\Sigma^T Q^T$$
$$= Q\Sigma I_m \Sigma^T Q^T$$
$$= Q\Sigma\Sigma^T Q^T,$$

we see that it constitutes a spectral decomposition of A by orthogonal matrices Q. This concludes the proof. □

It is often said that the PCA *chooses* the directions that contain the maximal variance of the data. To have this interpretation, the data must be *centered*. For a dataset $\{x_i\}_{i=1}^{m}$, we compute the mean vector

$$\mu = \frac{1}{m} \sum_{i=1}^{m} x_i,$$

and subtracting it from the datapoints, it yields the centered data

$$\tilde{x}_i = x_i - \mu \quad \forall i.$$

Let us write the sample covariance matrix for the centered data $\tilde{x}$, which is given by

$$C = \frac{1}{m} \sum_{i=1}^{m} (\tilde{x}_i - \tilde{\mu})(\tilde{x}_i - \tilde{\mu})^T = \frac{1}{m} \sum_{i=1}^{m} \tilde{x}_i \tilde{x}_i^T,$$

as $\tilde{\mu} = \frac{1}{m} \sum_{i=1}^{m} \tilde{x}_i = 0$. This corresponds to the matrix A for the centered data up to a constant $\frac{1}{m}$. Thus, in this case, the PCA finds the eigenvectors with largest corresponding eigenvalues, which are the directions where the covariance is largest. In Figure 10.2, we see an example comparing the PCA performed on uncentered, centered, and standardized data.

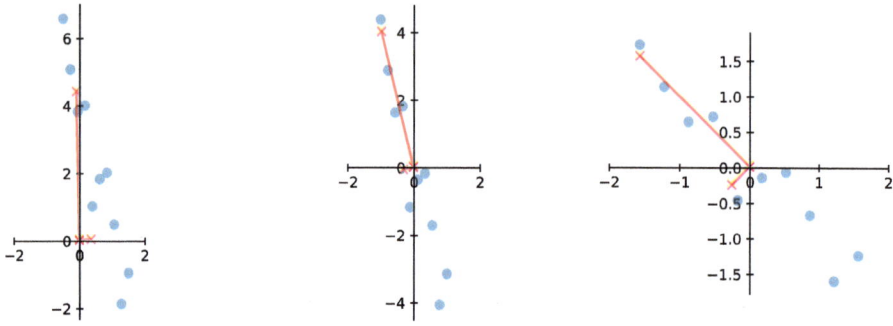

Figure 10.2: *Left:* PCA on uncentered data. *Middle:* PCA on centered data. *Right:* PCA on standardized data (zero mean and variance 1).

PCA has been applied to many problems since its conception. A popular example of PCA was to generate a *basis* for human faces, called eigenfaces, where human faces are represented in a compressed way. A picture of a human face in black and white can be represented as a matrix of pixel intensities $\mathcal{M} \in \mathbb{R}^{q \times q}$. Then such matrices can be flattened into vectors $x \in \mathbb{R}^n$, where $n = q \times q$. With these vectors, the eigenfaces can be computed. An example of this can be found in the implementation details at the end of this chapter.

10.2.2 Kernel PCA

The PCA method described previously finds linear maps W and U to perform the compression and reconstruction. We can extend it to nonlinear representations by using kernels. First, we will see that there is a way to write the PCA method to reveal a structure similar to the *kernel trick for SVMs*, and then we will define the Kernel PCA.

By Theorem 10.2.1 the solution to (10.4) is given by a matrix $U \in \mathbb{R}^{r \times n}$ that has r eigenvectors with the largest corresponding eigenvalues of A as its columns and $W = U^T$. Another way to find the eigenvalues of A is to solve the *eigenvalue problem* for the matrix A,

$$Av_j = \lambda_j v_j, \tag{10.6}$$

which finds an eigenvalue λ_j with the corresponding eigenvector v_j.

In this section, we first show that solving the eigenvalue problem (for nonzero eigenvalues) for A amounts to solving the eigenvalue problem to a related $m \times m$ matrix K:

$$K\vec{a}_j = \lambda_j \vec{a}_j, \tag{10.7}$$

where $K_{i,j} = x_i^T x_j$. Furthermore, $\vec{a}_j = (a_{j,1}, \dots, a_{j,m})$ is related to v_j through

$$v_j = \sum_{i=1}^{m} a_{j,i} x_i. \tag{10.8}$$

Lemma 10.2.4. *Eigenvectors with corresponding nonzero eigenvalues of A live in the span of the datapoints x.*

Proof. Let v be an eigenvector of A. Then

$$Av = \left(\sum_{i=1}^{m} x_i x_i^T \right) v = \sum_{i=1}^{m} (x_i^T v) x_i.$$

Since v can be written as a linear combination of the datapoints x_i, it is in the span of x. $\square$

Now we demonstrate the relation between (10.6) and (10.7). By Lemma 10.2.4 and (10.8), we can rewrite (10.6) as

$$Av_j = \sum_{i=1}^{m} \left(\sum_{k=1}^{m} a_{j,k} x_i^T x_k \right) x_i = \lambda_j \left(\sum_{k=1}^{m} a_{j,k} x_k \right).$$

Let x_ℓ be a datapoint in x. Multiplying the above by x_ℓ^T, we have

$$\sum_{i=1}^{m} \left(\sum_{k=1}^{m} a_{j,k} x_i^T x_k \right) x_\ell^T x_i = \lambda_j \left(\sum_{k=1}^{m} a_{j,k} x_\ell^T x_k \right), \quad \ell = 1, \ldots, m.$$

Notice that the equation above now depends on dot products of the data points and that the eigenvector v_j is determined by the coefficients $\vec{a}_j = (a_{j,1}, \ldots, a_{j,m})$.

Consider the matrix $K_{ij} = x_i^T x_j$. We can write the system of equations above simply as

$$K^2 \vec{a}_j = \lambda_j K \vec{a}_j.$$

Multiplying by K^{-1} yields

$$K \vec{a}_j = \lambda_j \vec{a}_j.$$

Solving the system above for $(\lambda_j, \vec{a}_j)$ is an eigenvalue problem for the matrix K, and $\vec{a}_j$ specifies the eigenvector v_j for matrix A.

Now that we wrote the problem with respect to K, we can consider other distance measures between elements, namely, any kernel we have studied before. This corresponds to considering the PCA of the data transformed under the corresponding map ϕ that sends point x to some RKHS. Abusing the notation, we now have the matrix A given by

$$A = \sum_{i=1}^{m} \phi(x_i) \phi(x_i)^T,$$

for which we find the principal components of. However, note that just like in kernel methods, we do not need to explicitly build this map.

Remark 10.2.5. To have a similar interpretation of the PCA picking the directions of maximal variation, we must have the centered transformed data, i. e.,

$$\frac{1}{m} \sum_{i=1}^{m} \phi(x_i) = 0.$$

10.3 Implementation details

In this section, we provide some implementation details for the PCA. On github, we provide a more extensive task using the well-known eigenface dataset [42].

Listing 10.1: PCA for a 2D dataset

```
 1 import numpy as np
 2 from sklearn.decomposition import PCA
 3 from sklearn.preprocessing import StandardScaler
 4
 5 # Generating 2D data
 6 X1 = np.linspace(0,2,10)-0.5
 7 X2 = -3.5*X1 + np.random.normal(0,1.0,size=(10))+4
 8
 9 # PCA on uncentered data
10 data_matrix = np.transpose(np.stack((X1,X2),axis=1))
11 U,S,V = np.linalg.svd(data_matrix)
12
13 pca_centered = PCA(n_components=2) #PCA centers the data
        by default
14 pca_centered.fit(data_matrix)
15 print(pca_scaled.components_)
16
17 scaler = StandardScaler()
18 data_matrix_standardised = scaler.fit_transform(
        data_matrix)
19 pca_scaled = PCA(n_components=2)
20 pca_scaled.fit(data_matrix_standardised)
21 print(pca_scaled.components_)
```

11 Reinforcement learning

So far, we have seen techniques to learn a function mapping from existing training data, that is, supervised learning, as well as to make sense of data through unsupervised methods such as clustering and dimensionality reduction.

Let us consider a new task. Consider a dynamical system that evolves over time. At a discrete time step n, the state of the system is described by an element s_n of some state space. A control u_n is applied, and the system then moves to a new state $s_{n+1} = f(n, s_n, u_n)$ for some transition map f. For a given sequence of states and controls (trajectory) of length N, the performance on this task is measured using a performance criterion $J((s_n, u_n)_{n \leq N})$.

Reinforcement learning (RL) is an approach to solve such problems through trials and errors: a learner (agent) samples a trajectory, collects *rewards* (i. e., an evaluation of its performance), and reinforces positively the chosen actions (controls) if the rewards are high and negatively if the rewards are low. It then repeats the procedure until it finds a satisfactory *policy* for choosing actions.

For example, the task of playing a chess game can be framed as a reinforcement learning task, where the agent plays a sequence of moves against an opponent to maximize the reward at the end of the game (which for example can be +1 for a win, –1 for a loss, and 0 for a draw). The type of tasks that reinforcement learning tackles are of sequential decision making and control problems. In this chapter, we introduce basic notions of reinforcement learning and general techniques used to solve these types of problems.

11.1 Markov decision processes

The setup for reinforcement learning can be informally summarized as follows: there exist an environment and a learner (or agent) in some state with respect to the environment. The agent performs an action and receives a reward, and the action makes the agent move to a different state. Over some sequential time, the agent seeks to maximize the cumulative reward through interactions with the environment by performing actions.

11.1.1 Setup

We assume that the agent interacts with the environment with a sequence of discrete time steps $t = 0, 1, 2, \ldots$, which can have a (possibly random) terminal time T, or continue indefinitely (finite horizon or infinite horizon). At each time step t the agent receives some representation of the environment state and its own internal state $S_t \in S$, and given that state, it selects an action $A_t \in \mathcal{A}$ and arrives at some other state S_{t+1}. In consequence of the tuple (S_t, A_t, S_{t+1}) of state, action, and next state, the agent receives

https://doi.org/10.1515/9783111288994-011

a reward $R_t \in \mathbb{R}$ and finds itself in a new state S_{t+1} (see Figure 11.1). This yields a sequence or *trajectory*[1]

$$S_0, A_0, R_0, S_1, A_1, R_2, S_2, A_2, \ldots .$$

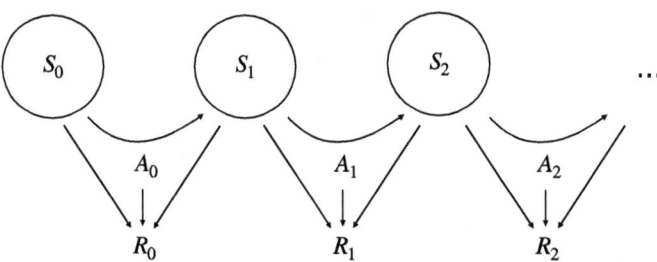

Figure 11.1: A schematic of the trajectory over time. The agent starts at state S_0 and performs action A_0, which leads it to state S_1. The reward for the tuple (S_0, A_0, S_1) is R_0.

At each time step t, we can define the *return at time t* as the sum of the current reward and all future rewards:

$$G_t := R_t + R_{t+1} + \cdots .$$

For infinite horizon tasks, it is convenient to define the *discounted return*

$$G_t := R_t + \gamma R_{t+1} + \gamma^2 R_{t+2} + \cdots$$

with $\gamma \in [0, 1]$, a parameter called the *discount rate*. This guarantees that the sum has a finite value if $\gamma < 1$ and $\{R_k\}_{k=0}^{\infty}$ are bounded. One way to interpret the discount rate is that it determines the present value of future rewards, namely, that a reward received in k steps ahead is only worth $\gamma^{k-1} r_{t+k}$.

The above dynamics describe the reaction of the environment (i. e., state transitions and rewards) to the agent's actions. But how do we proceed with decision making? We define a mapping, called a policy, that tells us what action to perform at any state.

Definition 11.1.1 (Policy). A policy is a mapping $\pi : \mathcal{A} \times \mathcal{S} \rightarrow [0, 1]$ such that

$$\pi(a|s) = P(A_t = a|S_t = s)$$

with the condition that $\sum_{a \in \mathcal{A}} \pi(a|s) = 1$ for all $s \in \mathcal{S}$. The policy map gives the probability of taking action a when the agent is in state s.

---

1 In this book, we adopt the following indexing for the sequence of states, actions, and rewards. However, it is also common to denote the first reward as R_1, and so on. Namely, for the tuple (S_t, A_t, S_{t+1}), the agent receives a reward R_{t+1}.

Remark 11.1.2. When $\pi(a|s) \in \{0,1\}$ for all $(a,s) \in \mathcal{A} \times \mathcal{S}$, we say that the policy is *deterministic*. A policy that is not deterministic is called *stochastic*, in which case the agent's actions are random variables.

From a stochastic policy we can define a deterministic policy that can be attained by computing the argmax:

$$a = \arg\max_{a \in \mathcal{A}} \pi(a|s).$$

This is called a *greedy policy*.

From a deterministic policy we can define the ϵ-greedy stochastic policy as follows: at each step, with probability $1 - \epsilon$, follow the greedy policy and with probability ϵ, choose an action uniformly at random.

We now formalize the dynamics of the environment.

Definition 11.1.3 (Markov decision process). Let the tuple $(\mathcal{S}, \mathcal{A}, p, r)$ be given as follows:
- $\mathcal{S}$ is a finite set of states;
- $\mathcal{A}$ is a finite set of actions;
- p is the state transition kernel $p(s, a, s')$, the conditional probability that the agent transits to s' given that it takes action a at state s; and
- $r : \mathcal{S} \times \mathcal{A} \times \mathcal{S} \to \mathbb{R}$ is a reward function.

Given a policy π and an initial state distribution v on $\mathcal{S}$, the MDP is constructed recursively as follows:
- At time $t = 0$ the agent is located at $S_0 \sim v$.
- At state S_t, the agent takes action $A_t \sim \pi(\cdot|S_t)$, transits to state $S_{t+1} \sim p(S_t, A_t, \cdot)$, and collects a reward $R_t = r(S_t, A_t, S_{t+1})$, independently from all past random variables.

We make two simplifying assumptions on the reward to ease the presentation:
1. In full generality, the reward can be a random variable with conditional law $\mu_R(\cdot, S_t, A_t, S_{t+1})$. We assume that it is generated from a deterministic function in this chapter.
2. We will moreover assume that the reward function does not depend on the next state, so that $R_t = r(A_t, S_t)$, and that it is bounded, i. e., $\|r\|_\infty < \infty$.

Thanks to the recursive definition, we see that the process of states and actions satisfy the so-called *Markov property* (hence the name Markov decision process):

$$\mathbb{P}\big[(A_{t+k}, S_{t+k+1})_{k \geq 0} = (a_{t+k}, s_{t+k+1})_{k \geq 0} | (S_{j+1}, A_j)_{j \leq t-1}\big]$$
$$= \mathbb{P}\big[(A_{t+k}, S_{t+k+1})_{k \geq 0} = (a_{t+k}, s_{t+k+1})_{k \geq 0} | S_t\big]. \tag{11.1}$$

Note that we have the analogous statement when additionally conditioning on A_t. In other words, the Markov property tells us that at any time t, the only piece of information that is relevant in the history of the MDP to characterize the law of its future is contained in the current state S_t.

11.1.2 Value function and Q-function

Depending on the policy of the agent, a state can be evaluated according to the expectation of the future rewards. Similarly, we can evaluate an action at a given state. We can naturally define two important functions, the *state value function* and the *state-action value function*.

Definition 11.1.4 (State value function for policy π). The state value function $v_\pi(s)$ (often called just value function) at some time t gives the *value* of a state s under a policy π, defined by the expected return starting at state s and following policy π afterwards:

$$v_\pi(s) := \mathbb{E}_\pi[G_t|S_t = s] = \mathbb{E}_\pi\left[\sum_{k=0}^\infty \gamma^k R_{t+k}|S_t = s\right], \tag{11.2}$$

where $\mathbb{E}_\pi[\cdot]$ denotes the expected value along a random trajectory generated under π.

The value of a terminal state is always zero. Note that by the Markov property the value function does not depend on t, so we might also take $t = 0$.

Similarly to the state value function for policy π, the state-action value function (often called just action value function or Q-function) yields the expected return starting from state s, taking action a, and then following policy π.

Definition 11.1.5 (State-action value function). The state-action value function q_π is defined for all $t \geq 0$ by

$$q_\pi(a, s) := \mathbb{E}_\pi[G_0|S_0 = s, A_0 = a] = \mathbb{E}_\pi\left[\sum_{k=0}^\infty \gamma^k R_k|S_0 = s, A_0 = a\right].$$

As for the state value function, q_π does not depend on t.

Remark 11.1.6. In general, if the state-action space is large or if the dynamics is unknown, the functions v_π and q_π cannot be computed and are therefore estimated from experience. For example, if an agent follows policy π and maintains an average of the returns obtained from any given state that has been encountered, then this average converges (provided that the MDP is irreducible and ergodic, which is the case for finite state and action spaces where all states are reachable from any other) to the state value $v_\pi(s)$ as the number of times each state is encountered approaches infinity. Similarly, if the average is kept for each action a and s, then this average converges to $q_\pi(s, a)$.

There is clearly a relation between a policy and its value functions. We will see in this chapter that one popular way to solve RL problems is to learn the optimal policy by using these value functions. However, before we focus on the following theorem, which is a fundamental result for reinforcement learning, as it expresses a relationship between the value of a state and the values of its successor states.

Theorem 11.1.7 (Consistency condition for v_π). *For any policy π and any state s, the following consistency condition holds between the value function at s and the value of its possible successor states s':*

$$v_\pi(s) = \sum_{a\in\mathcal{A}} \pi(a|s)\left(r(a,s) + \gamma \sum_{s'\in\mathcal{S}} p(s,a,s')v_\pi(s')\right) \quad \forall s \in \mathcal{S}. \tag{11.3}$$

Note that we can alternatively write this as

$$v_\pi(s) = \mathbb{E}_\pi[R_t + \gamma V_\pi(S_{t+1})|S_t = s].$$

Proof. Firstly, by definition, for any arbitrary $t \geq 0$, we have that $G_t = R_t + \gamma G_{t+1}$. By the definition of v_π and the Markov property we have that $\mathbb{E}_\pi[G_{t+1}|S_{t+1}, S_t] = v_\pi(S_{t+1})$. The tower property of conditional expectation yields that

$$\begin{aligned}
v_\pi(s) &= \mathbb{E}_\pi[G_t|S_t = s] \\
&= \mathbb{E}_\pi[R_t + \gamma G_{t+1}|S_t = s] \\
&= \mathbb{E}_\pi[R_t + \gamma\mathbb{E}[G_{t+1}|S_{t+1}, S_t = s]|S_t = s] \\
&= \mathbb{E}_\pi[R_t + \gamma v_\pi(S_{t+1})|S_t = s].
\end{aligned}$$

Then it suffices to use the definition of the expectation: the conditional joint law of (A_t, S_{t+1}) given $S_t = s$ is given in the construction of the MDP, and we can write

$$\begin{aligned}
v_\pi(s) &= \sum_{a\in\mathcal{A}}\sum_{s'\in\mathcal{S}} \pi(a|s)p(s,a,s')(r(a,s) + \gamma v_\pi(s')) \\
&= \sum_{a\in\mathcal{A}} \pi(a|s)\left(r(a,s) + \sum_{s'\in\mathcal{S}} p(s,a,s')\gamma v_\pi(s')\right),
\end{aligned}$$

where we used that $\sum_{s'\in\mathcal{S}} p(s,a,s') = 1$ for all a, s. This concludes the proof. $\quad\square$

The consistency relation (11.3), also called the *Bellman equation for* v_π, states that the value function at s is given by a contribution of the immediate reward plus the discounted value function at the next state s'. For any given policy π, the value function v_π is the unique solution to (11.3), and thus we can use this fact to find the value function for any policy π.

11.2 Optimality

In this section, we discuss two main approaches to solve the MDP presented above, which means finding a policy that maximizes the cumulative return.

We say that a policy π is better than or equal to another policy π' if

$$v_\pi(s) \geq v_{\pi'}(s) \quad \forall s \in \mathcal{S}.$$

Note that this is only a partial order, and a policy π could be neither better nor worse than another policy π'. However, there is always at least one policy that is better than or equal to all other policies. This is called an optimal policy π^*, which does not need to be unique. However, all optimal policies share the same value function. Namely, the optimal state value function is defined as

$$v^*(s) := \max_\pi v_\pi(s) \quad \forall s \in \mathcal{S}, \tag{11.4}$$

whereas the optimal action-state value function is defined as

$$q^*(a, s) := \max_\pi q_\pi(a, s) \quad \forall s \in \mathcal{S}, \forall a \in \mathcal{A}. \tag{11.5}$$

For a deterministic policy π, we write $\pi(s)$ for the action a such that $\pi(a|s) = 1$. A consequence from (11.4) and (11.5) is that we can obtain a deterministic optimal policy as

$$\pi^*(s) = 1 \quad \text{if and only if} \quad \pi^*(s) = \arg\max_{a \in \mathcal{A}} q^*(a, s).$$

Theorem 11.2.1 (Bellman's optimality equations). *The optimal state value and action-state value functions are unique and related by*

$$v^*(s) = \max_{a \in \mathcal{A}} q^*(s, a), \tag{11.6}$$

$$q^*(s, a) = \mathbb{E}_{\pi^*}\left[R_t + \gamma v^*(S_{t+1})|S_t = s, A_t = a\right]. \tag{11.7}$$

11.3 Iterative methods

The key idea in this section is that we can use the value functions to find good policies. Firstly, for a given policy, we need to compute the corresponding value function, which can be done thanks to Theorem 11.1.7. Note that if the transition probability kernel and reward r are known, then we have a linear system of $|\mathcal{S}|$ equations with $|\mathcal{S}|$ unknowns, which we can solve explicitly. Let $\mathcal{S} = \{s_1, \ldots, s_{|\mathcal{S}|}\}$. Then we have in matrix form

$$\vec{v}_\pi = \vec{R} + \gamma P \vec{v}_\pi \tag{11.8}$$

with

$$\vec{v}_{\pi,i} = v_\pi(s_i), \quad \vec{R}_i = \sum_{a \in \mathcal{A}} \pi(a|s_i) r(a, s_i), \quad P_{i,j} = \sum_{a \in \mathcal{A}} \pi(a|s_i) p(s_i, a, s_j).$$

Solving this linear system directly can be cumbersome (complexity of $\mathcal{O}(|\mathcal{S}|^3)$). Thus we can solve for $\vec{v}_\pi$ in an iterative fashion:

$$\vec{v}_\pi^{(k+1)} = \vec{R} + \gamma P \vec{v}_\pi^{(k)}, \tag{11.9}$$

given some initial guess, e. g., setting $\vec{v}_\pi^{(0)} = \vec{0}$.

We show that the iterative procedure (11.9) converges to the unique solution of (11.8). For this, we use the Banach fixed point theorem.[2] Then we can prove the convergence of the iterative process (11.9).

Theorem 11.3.2 (Convergence of the iterative Bellman equation). *For any policy π, the iterative process (11.9) converges to the corresponding value function v_π. Moreover, the rate of convergence is given by*

$$\left\| v^{(k)} - v_\pi \right\|_\infty \leq \gamma^k \left\| v^{(0)} - v_\pi \right\|_\infty.$$

Proof. Let π be a fixed policy and define the map $T(v) = \vec{R} + \gamma P v$ on $\mathbb{R}^{|S|}$. We show this map is a contraction with respect to the infinity norm. Letting $v, v' \in \mathbb{R}^{|S|}$, we have

$$\left\| T(v) - T(v') \right\|_\infty = \left\| \gamma(Pv - Pv') \right\|_\infty = \gamma \left\| P(v - v') \right\|_\infty \leq \gamma \|P\|_\infty \|v - v'\|_\infty,$$

where $\|P\|_\infty := \sup_{\|v\|_\infty = 1} \|Pv\|_\infty$. Note that $\|P\|_\infty \leq 1$ because P is a stochastic matrix. Therefore we have

$$\left\| T(v) - T(v') \right\|_\infty \leq \gamma \|v - v'\|_\infty,$$

which shows that T is a contraction. By the Banach fixed point theorem, we have that T has a unique fixed point $v \in \mathbb{R}^{|S|}$, and iteration (11.9) converges with the corresponding convergence rate. To conclude, it suffices to recall that v_π is a fixed point of T by (11.8). $\qquad\square$

By the above theorem, for a given policy π, we can approximate v_π. How can we then find an optimal policy? Once we have v_π, we select some action a at state s and then follow the existing policy π. We thus compute the action value function $q_\pi(s, a)$. If $q_\pi(s, a) > v_\pi(s)$ for some $a \in \mathcal{A}$, then this means that at state s, it is better to take action a and only then follow π, instead of following π at all states. We can formalize this in the next theorem.

Theorem 11.3.3 (Policy improvement theorem). *Consider two deterministic policies π and π'. If $q_\pi(s, \pi'(s)) \geq v_\pi(s)$ for all $s \in \mathcal{S}$, then $v_{\pi'}(s) \geq v_\pi(s)$ for all $s \in \mathcal{S}$.*

---

2 **Theorem 11.3.1** (Banach's fixed point theorem [9]). *If the function T is a contraction on the complete metric space (X, d) with contraction constant γ, then T has a unique fixed point $x^* \in X$. Moreover, if $x \in X$, then the sequence $\{T^n(x)\}_{n=0}^\infty$ converges to x^* as $n \to \infty$, and*

$$d\left(T^n(x), x_0\right) \leq \frac{\gamma^n}{1 - \gamma} d(x, x_0).$$

Proof. We write

$$v_\pi(s) \le q_\pi(s, \pi'(s))$$
$$= \mathbb{E}[R_t + \gamma v_\pi(S_{t+1})|S_t = s, A_t = \pi'(s)]$$
$$= \mathbb{E}_{\pi'}[R_t + \gamma v_\pi(S_{t+1})|S_t = s] \quad (\pi' \text{ is deterministic})$$
$$\le \mathbb{E}_{\pi'}[R_t + \gamma q(S_{t+1}, \pi'(S_{t+1}))|S_t = s]$$
$$= \mathbb{E}_{\pi'}[R_t + \gamma(R_{t+1} + \gamma v_\pi(S_{t+2}))|S_t = s]$$
$$\le \mathbb{E}_{\pi'}[R_t + \gamma(R_{t+1} + \gamma q(S_{t+2}, \pi'(S_{t+2})))|S_t = s]$$
$$\le \cdots \quad \text{(recursively using the definition of } q \text{ and inequality)}$$
$$\le \mathbb{E}_{\pi'}[R_t + \gamma R_{t+1} + \gamma^2 R_{t+2} + \gamma^3 R_{t+3} + \cdots |S_t = s]$$
$$= v_{\pi'}(s). \qquad \square$$

Using Theorem 11.3.3, we can write a greedy algorithm, called *policy improvement*, to update the policy π:

$$\pi^{(k+1)}(s) = \arg\max_a q_{\pi^{(k)}}(a, s) \quad \forall s \in \mathcal{S}. \tag{11.10}$$

Suppose that the new policy $\pi^{(k+1)}$ is as good as (but not better than) the previous policy $\pi^{(k)}$. This means that $v_{\pi^{(k+1)}} = v_{\pi^{(k)}}$. By (11.10) we have that

$$v_{\pi^{(k)}}(s) = \max_a q_{\pi^{(k)}}(a, s) \quad \forall s \in \mathcal{S},$$

which corresponds to the Bellman optimality equation (11.6). Therefore $v_{\pi'}$ must be v_*, and both π and π' must be optimal policies.

We can also directly find the optimal value function and then derive the policy from that. Using (11.6), we can write the recursion

$$v^{(k+1)}(s) = \max_{a \in \mathcal{A}} \sum_{s' \in \mathcal{S}} [r(a, s) + \gamma p(s, a, s')v^{(k)}(s')] \quad \forall s \in \mathcal{S}.$$

For arbitrary initial conditions, the sequence $\{v_k\}$ converges to v_* (under the same conditions that guarantee the existence of v_*).

Remark 11.3.4. Note that for both iterative methods introduced, we must either know or have some estimate of the transition probabilities $p(s, a, s')$ to be able to derive the value functions and therefore policies. This is called *model-based reinforcement learning*, which requires to know or learn the transition probabilities between states (see [33] for a comprehensive survey).

When $p(s, a, s')$ is unknown, we can use Monte Carlo approximations based on sampled trajectories. Such algorithms are then called *model free* as they approximate q_π and v_π without learning the transition probabilities $p(s, a, s')$.

11.4 Policy gradient method

Although the value-function approach has worked well in different applications, it has some limitations, such as computational complexity when the state and action spaces are large. Furthermore, it is tailored toward finding deterministic policies, and small variations in the value-function can lead to very different policies. In contrast to the methods presented in the previous section, in what follows, we consider a parametric policy that is updated directly, based on its performances. The general algorithm is called *policy gradient*; the name will become clear later on.

11.4.1 Parametric objective

Henceforth, we consider a parametric policy π_θ where $\theta \in \mathbb{R}^P$ is a vector of parameters. For now, we stay agnostic to the specific parametric model that is being used, and we write θ for the vector of parameters. Learning a policy thus means learning optimal parameters θ of the model π_θ. The first problem we encounter is that the notion of optimal policy is less clear in the parametric case. To see why, consider $\mathcal{S} = \{0, 1\}$ and $\mathcal{A} = \{0, 1\}$. The transition probability kernel is simply $p(0, a, 1) = p(1, a, 0) = 1$ for both $a = 0$ and $a = 1$, that is, no matter what the agent does, it always goes from state 0 to state 1 and from state 1 to state 0. The reward function is the following: $r(0, 0) = r(0, 1) = 1$, and $r(1, 0) = r(1, 1) = 0$. Consider the parameter $\theta \in \mathbb{R}$ and the parametric policy $\pi_\theta(a|s) = \mathbb{1}_{\{a=s\}}\mathbb{1}_{\{\theta\geq0\}} + \mathbb{1}_{\{a\neq s\}}\mathbb{1}_{\{\theta<0\}}$. We see that for any fixed θ, $\pi_\theta(0|0) = \pi_\theta(1|1)$ and $\pi_\theta(0|1) = \pi_\theta(1|0)$. We can check that for $\theta_+ \geq 0$ and $\theta_- < 0$, we get $v_{\pi_{\theta_+}}(0) > v_{\pi_{\theta_-}}(0)$, but $v_{\pi_{\theta_+}}(1) < v_{\pi_{\theta_-}}(1)$. Therefore no *parametric* policy dominates all others.

There is nonetheless a natural way to define an objective function: let v be the *initial state distribution*, that is, a probability on $\mathcal{S}$ such that the initial state $S_0 \sim v$. The goal of the agent is therefore to maximize the following objective function:

$$J(\theta) = \sum_{s \in \mathcal{S}} v_{\pi_\theta}(s)v(s). \tag{11.11}$$

Note that this can be rewritten $J(\theta) = \mathbb{E}_\pi[\sum_{k\geq0} \gamma^k r(A_k, S_k)]$, where $S_0 \overset{d}{\sim} v$, which is the expected cumulative reward of the agent starting from an initial state with law v.

As in the simple example above, there is no guarantee that there exists θ_* such that value function $v_{\pi_{\theta_*}}$ dominates v_{π_θ} for all θ, nor $\arg\max_\theta J(\theta)$ is unique, that is, several policies can perform equally well in average. Moreover, a different initial state distribution v' can lead to different optimal policies.

11.4.2 Policy gradient theorem

As opposed to iterative methods optimizing the value function or the Q-function, policy gradient (PG) methods improve the policy directly by positively reinforcing actions that lead to positive rewards and negatively reinforcing those that lead to negative rewards. Recall that the goal of the agent is to maximize $J(\theta)$ defined in (11.11). We assume that the parameterization $\theta \mapsto \pi_\theta$ is differentiable, and hence we would like to update the policy parameters through a gradient ascent on J:

$$\theta_{t+1} = \theta_t + \eta \nabla J(\theta), \tag{11.12}$$

where η is a learning rate.

Note that computing the gradient of $J(\theta)$ seems rather complicated since π_θ appears infinitely many times in $v_{\pi_\theta}(s) = \sum_{k\geq 1} \gamma^k \mathbb{E}_\pi[r(A_k, S_k)|S_0 = s]$, where, moreover, S_k is sampled after choosing k actions under π. However, to take advantage of the Markov property of MDPs, we first need some definitions.

Let $\theta \in \mathbb{R}^P$, and let $P(\theta) \in \mathbb{R}^{|S|\times|S|}$ be the matrix defined by $P(\theta)_{s,s'} :=$ $\sum_{a\in A} \pi_\theta(a|s)p(s, a, s')$. The value of $P(\theta)_{s,s'}$ is the probability that the agent goes from state s to state s' in one step following policy π_θ. More generally, raised to the power $n \geq 1$, the matrix $P(\theta)^n_{s,s'}$ is the probability that the agent goes from s to s' in n steps, that is, $P(\theta)^n_{s,s'} = \mathbb{P}_{\pi_\theta}[S_n = s|S_0 = s']$, which can easily be proved using the Markov property.

Lemma 11.4.1 (Average visits to states). *Let $\rho_{\pi_\theta}(s, s')$ be the expected discounted number of visits of state s' by the agent during one episode, starting at s and following policy π_θ, that is,*

$$\rho_{\pi_\theta}(s, s') = \mathbb{E}_{\pi_\theta}\left[\sum_{k\geq 0} \gamma^k \mathbb{1}_{\{S_k=s'\}}|S_0 = s\right].$$

Then we have

$$\rho_{\pi_\theta}(s, s') = \sum_{k\geq 0} \gamma^k P(\theta)^k_{s,s'}.$$

Proof. The proof is straightforward from the definition of the matrix $P(\theta)$. We simply write

$$\rho_{\pi_\theta}(s, s') = \mathbb{E}_{\pi_\theta}\left[\sum_{k\geq 0} \gamma^k \mathbb{1}_{\{S_k=s'\}}|S_0 = s\right]$$

$$= \sum_{k\geq 0} \gamma^k \mathbb{E}_{\pi_\theta}[\mathbb{1}_{\{S_k=s'\}}|S_0 = s],$$

where we could take the sum out of the expectation by the monotone convergence theorem since all terms are nonnegative, and then by the definition of the expectation, $\mathbb{E}_{\pi_\theta}[\mathbb{1}_{\{S_k=s'\}}|S_0 = s] = \mathbb{P}_{\pi_\theta}[S_k = s'|S_0 = s]$, which proves the claim. □

The next theorem enables us to compute the gradient of J.

Theorem 11.4.2 (Policy gradient theorem). *The gradient of the objective $J(\theta)$ with respect to θ is given by*

$$\nabla J(\theta) = \sum_{s_0 \in S} v(s_0) \sum_{s \in S} \rho_{\pi_\theta}(s_0, s) \sum_{a \in A} q_{\pi_\theta}(a, s) \nabla \pi_\theta(a|s).$$

This theorem provides an analytic expression for the gradient of the objective with respect to the policy parameters that only depends on the gradient of the policy and not on that of the discounted average state distribution ρ_{π_θ}, which makes the computations tractable. To compute the gradient, we only need $q_\pi(s, a)$. In practice, the gradient is approximated using a Monte Carlo estimate of $q_\pi(s, a)$. We can proceed to prove Theorem 11.4.2.

Proof. We write

$$\nabla v_{\pi_\theta}(s) = \nabla \left[\sum_{a \in A} \pi_\theta(a|s) q_{\pi_\theta}(a, s) \right]$$

$$= \underbrace{\sum_{a \in A} \nabla \pi_\theta(a|s) q_{\pi_\theta}(a, s)}_{\phi(s) :=} + \sum_{a \in A} \pi_\theta(a|s) \nabla q_{\pi_\theta}(a, s)$$

$$= \phi(s) + \sum_{a \in A} \pi_\theta(a|s) \nabla \left[r(a, s) + \sum_{s' \in S} \gamma p(s, a, s') v_{\pi_\theta}(s') \right]$$

$$= \phi(s) + 0 + \sum_{s' \in S} \gamma P(\theta)_{s,s'} \nabla v_{\pi_\theta}(s').$$

This gives a recursion for ∇v_{π_θ}. Applying it to the right-hand side yields

$$\nabla v_{\pi_\theta}(s) = \phi(s) + \sum_{s' \in S} \gamma P(\theta)_{s,s'} \left[\phi(s') + \sum_{s'' \in S} \gamma P(\theta)_{s',s''} \nabla v_{\pi_\theta}(s'') \right]$$

$$= \phi(s) + \sum_{s' \in S} \gamma P(\theta)_{s,s'} \phi(s') + \sum_{s'' \in S} \gamma^2 P(\theta)^2_{s,s''} \nabla v_{\pi_\theta}(s'').$$

Writing out the recursion, we have

$$\nabla v_{\pi_\theta}(s) = \sum_{s' \in S} \sum_{k \geq 0} \gamma^k P(\theta)^k_{s,s'} \phi(s')$$

$$= \sum_{s' \in S} \rho_{\pi_\theta}(s, s') \phi(s').$$

We now sum over the initial states to get

$$\sum_{s_0 \in S} v(s_0) \nabla v_{\pi_\theta}(s_0) = \sum_{s_0 \in S} v(s_0) \sum_{s \in S} \rho_{\pi_\theta}(s_0, s) \phi(s).$$

Substituting $\phi(s)$ as defined previously, we obtain the result as claimed, which concludes the proof. $\square$

11.4.3 Softmax policies

The policy can be represented by any of the models we have discussed in this book, such as linear models, kernel-based models, ensemble models, neural networks, etc., followed by transformation yielding a probability distribution. A convenient choice is to apply a *softmax* transformation defined as follows: let $h \in \mathbb{R}^d$ for some $d \geq 2$; then the softmax transformation σ of the vector h is given by

$$\sigma_\tau(h) = \frac{1}{Z} \begin{pmatrix} e^{h_1} \\ \vdots \\ e^{h_d} \end{pmatrix},$$

$$\text{where} \quad Z = \sum_{i=1}^{d} e^{h_i}.$$

By definition, $\sigma(h)$ is always a probability vector, that is, $\sigma(h_i) \in (0, 1)$ for all $i = 1, \ldots, d$, and $\sum_{i=1}^{d} S(h)_i = 1$.

The name *softmax* comes from the fact that the components on which $\sigma(h)$ concentrates the most are the larger ones, but the probability thus obtained always gives positive mass to all components: it is softer than the arg max, and its maximal component is the arg max. We call the components of h the *action preferences* of the softmax probability $\sigma(h)$.

The softmax is a very natural way to obtain a probability from an arbitrary vector. We refer the interested reader to [21] for a more general definition of the softmax transformation and its properties, as well as further motivations for its use in RL.

To parameterize the softmax policy, consider a linear model for the action preference

$$h_\theta(a, s) := \theta \cdot \psi(a, s)$$

for some feature map $\psi : \mathcal{A} \times \mathcal{S} \to \mathbb{R}^d$, $d \geq 1$. Such policies are sometimes called *log-linear policies*.

It is an easy exercise to show that the gradient of the logarithm of the policy with respect to the parameters θ is given at any $(a, s) \in \mathcal{A} \times \mathcal{S}$ by

$$\nabla \log \pi_\theta(a|s) = \nabla h_\theta(a, s) - \sum_{a' \in \mathcal{A}} \pi_\theta(a'|s) \nabla h_\theta(a', s)$$

$$= \psi(a, s) - \mathbb{E}_{\pi_\theta}[\psi(A, s)]. \tag{11.13}$$

The next proposition gathers some useful properties of log-linear policies.

Proposition 11.4.3. *Suppose that $\|r\|_\infty < \infty$ and that there exists $C_\psi > 0$ such that $\|\psi(a, s)\|_2 < C_\psi$ for all $s \in \mathcal{S}$ and $a \in \mathcal{A}$. Then there exists $C > 0$ such that for all $\theta_1, \theta_2 \in \mathbb{R}^P$, $s \in \mathcal{S}$, and $a \in \mathcal{A}$, we have*

$$\left|\pi_{\theta_1}(a|s) - \pi_{\theta_2}(a,s)\right| \le C\|\theta_1 - \theta_2\|_2, \qquad \text{(i)}$$

$$\left|\nabla\pi_{\theta_1}(a|s) - \nabla\pi_{\theta_2}(a,s)\right| \le C\|\theta_1 - \theta_2\|_2, \qquad \text{(ii)}$$

$$\left|q_{\pi_{\theta_1}}(a,s) - q_{\pi_{\theta_2}}(a,s)\right| \le C\|\theta_1 - \theta_2\|_2, \qquad \text{(iii)}$$

$$\left|\rho_{\pi_{\theta_1}}(a,s) - \rho_{\pi_{\theta_2}}(a,s)\right| \le C\|\theta_1 - \theta_2\|_2, \qquad \text{(iv)}$$

and furthermore, $\|\nabla\log\pi_{\theta_1}(a|s)\|_2 \le C$ *for all* $s \in \mathcal{S}$ *and* $a \in \mathcal{A}$.

Proof. **(i)** A Taylor expansion of the parameterization yields that there exists ϑ on the d-dimensional segment $[\theta_1, \theta_2]$ such that

$$\left|\pi_{\theta_1}(a|s) - \pi_{\theta_2}(a|s)\right| = \left|(\theta_1 - \theta_2) \cdot \nabla\pi_\vartheta(a|s)\right|$$
$$\le \|\theta_1 - \theta_2\|_2 \|\nabla\pi_\vartheta(a|s)\|_2$$

by Cauchy–Schwarz inequality. By (11.13) and the boundedness assumption on ψ we have that

$$\|\nabla\pi_\vartheta(a|s)\|_2 \le \sup_{a' \in \mathcal{A}} 2\|\psi(a',s)\|_2 \le 2C_\psi, \qquad (11.14)$$

which proves the first claim.

(ii) We use (11.13) to write

$$\left\|\nabla\pi_{\theta_1}(a|s) - \nabla\pi_{\theta_2}(a|s)\right\|_2^2$$
$$= \pi_{\theta_1}(a|s)(\psi(a,s) - \mathbb{E}_{\pi_{\theta_1}}[\psi(A,s)]) \cdot \pi_{\theta_2}(a|s)(\psi(a,s) - \mathbb{E}_{\pi_{\theta_2}}[\psi(A,s)])$$
$$= \pi_{\theta_1}(a|s)\mathbb{E}_{\pi_{\theta_1} \otimes \pi_{\theta_2}}[(\psi(a,s) - \psi(A,s)) \cdot \pi_{\theta_2}(a|s)(\psi(a,s) - \psi(A',s))]$$
$$\le \mathbb{E}_{\pi_{\theta_1}}\left[\|\psi(a,s) - \psi(A,s)\|^2\right]\mathbb{E}_{\pi_{\theta_2}}\left[\|\psi(a,s) - \psi(A,s)\|^2\right]$$

by the Cauchy–Schwarz inequality, where $A \sim \pi_{\theta_1}(\cdot|s)$ and $A' \sim \pi_{\theta_2}(\cdot|s)$ are independent. Thanks to the boundedness of feature maps, we obtain

$$\left\|\nabla\pi_{\theta_1}(a|s) - \nabla\pi_{\theta_2}(a|s)\right\|_2^2 \le 4C_\psi^2,$$

which proves the second claim.

(iii) Note that since the rewards are uniformly bounded, by the definition of q_{π_θ} we have that $\|q_\pi\|_\infty \le \sum_{k\ge 0}\gamma^k\|r\|_\infty \le \frac{\|r\|_\infty}{1-\gamma}$. We first write

$$\left|v_{\pi_{\theta_1}}(s) - v_{\pi_{\theta_2}}(s)\right|$$
$$= \left|\sum_{a \in \mathcal{A}}(\pi_{\theta_1}(a|s)q_{\pi_{\theta_1}}(a,s) - \pi_{\theta_2}(a,s)q_{\pi_{\theta_2}}(a,s))\right|$$
$$\le \left|\sum_{a \in \mathcal{A}}(\pi_{\theta_1}(a|s) - \pi_{\theta_2}(a,s))q_{\pi_{\theta_1}}(a,s)\right| + \left|\sum_{a \in \mathcal{A}}\pi_{\theta_2}(a,s)(q_{\pi_{\theta_1}}(a,s) - q_{\pi_{\theta_2}}(a,s))\right|$$

$$\leq \|q_{\pi_{\theta_1}}\|_\infty \sum_{a\in\mathcal{A}} |\pi_{\theta_1}(a|s) - \pi_{\theta_2}(a|s)| + \sum_{a\in\mathcal{A}} |q_{\pi_{\theta_1}}(a|s) - q_{\pi_{\theta_2}}(a|s)|$$

$$\leq \frac{\|r\|_\infty}{1-\gamma} C\|\theta_1 - \theta_2\|_2 + \sum_{a\in\mathcal{A}} |q_{\pi_{\theta_1}}(a|s) - q_{\pi_{\theta_2}}(a|s)|,$$

where we used the first claim of the proposition. To prove the bound for the action-state function, it now suffices to use the above computation to note that

$$|q_{\pi_\theta}(a,s) - q_{\pi_{\theta'}}(a,s)| = \gamma \sum_{s'\in\mathcal{S}} p(s,a,s')|V_{\pi_\theta}(s') - V_{\pi_{\theta'}}(s')|$$

$$\leq \gamma \frac{\|r\|_\infty}{1-\gamma} C\|\theta_1 - \theta_2\|_2 + \gamma \sum_{s'\in\mathcal{S}} p(s,a,s') \sum_{a'\in\mathcal{A}} |q_{\pi_{\theta_1}}(a'|s') - q_{\pi_{\theta_2}}(a'|s')|.$$

By induction we obtain

$$|q_{\pi_\theta}(a,s) - q_{\pi_{\theta'}}(a,s)| \leq \frac{\|r\|_\infty}{(1-\gamma)^2} C\|\theta_1 - \theta_2\|_2,$$

which proves the third claim.

(iv) By Lemma 11.4.1 we have

$$|p_{\pi_{\theta_1}}(s,s') - p_{\pi_{\theta_2}}(s,s')| \leq \sum_{k\geq 0} \gamma^k |P(\theta_1)^k_{s,s'} - P(\theta_2)^k_{s,s'}|$$

$$\leq \sum_{k\geq 0} \gamma^k \sum_{s_i,a_i} \prod_{i=0}^{k-1} p(s_i,a_i,s_{i+1}) |\pi_{\theta_1}(a_i|s_i)^k - \pi_{\theta_2}(a_i|s_i)^k|,$$

where for every term k in the sum, the state sequence must satisfy $s_0 = s$ and $s_k = s$. The term $k = 0$ is easily bounded by $C\|\theta_1 - \theta_2\|_2$ as in (i). For the terms $k \geq 1$, we factor out $\pi_{\theta_1}(a_0|s) - \pi_{\theta_2}(a_0|s)$ and upper bound the remaining by 1, since it is a transition probability. This yields the upper bound

$$|p_{\pi_{\theta_1}}(s,s') - p_{\pi_{\theta_2}}(s,s')| \leq C\|\theta_1 - \theta_2\|_2 + \sum_{a_0} |\pi_{\theta_1}(a_0|s) - \pi_{\theta_2}(a_0|s)| \sum_{k\geq 1} \gamma^k$$

$$\leq C\|\theta_1 - \theta_2\|,$$

where we used (i), and where the constant C changed from one line to the other and is independent from s, s', which yields the claim.

The final claim is easy to see from (11.13) and the boundedness of ψ, which concludes the proof. □

Theorem 11.4.4. *For the log-linear parameterization, there exists $\eta_0 > 0$ such that for all $\eta \in (0,\eta_0)$, the objective $J(\theta_k)$ trained with the policy gradient algorithm of (11.12) converges, that is, $\lim_{k\to\infty} J(\theta_k)$ exists.*

It is important to note that since the optimal policy is deterministic and the softmax policy can only express stochastic policies for finite θ, it can be the case that θ_k does not converge even though $J(\theta_k)$ does.

Proof. To prove the claim, it suffices to show that the objective J is L-smooth for some $L > 0$ and then apply Theorem 3.2.4. Let $\theta_1, \theta_2 \in \mathbb{R}^P$ be two vectors of parameters. The policy gradient theorem (Theorem 11.4.2) entails that

$$
\nabla J(\theta_1) - \nabla J(\theta_2) = \sum_{s_0 \in \mathcal{S}} v(s_0) \sum_{s \in \mathcal{S}} \Big(p_{\pi_{\theta_1}}(s_0, s) \sum_{a \in \mathcal{A}} q_{\pi_{\theta_1}}(a, s) \nabla \pi_{\theta_1}(a|s)
$$
$$
- p_{\pi_{\theta_2}}(s_0, s) \sum_{a \in \mathcal{A}} q_{\pi_{\theta_2}}(a, s) \nabla \pi_{\theta_2}(a|s) \Big)
$$
$$
= \sum_{s_0 \in \mathcal{S}} v(s_0) \sum_{s \in \mathcal{S}} \Big((p_{\pi_{\theta_1}}(s_0, s) - p_{\pi_{\theta_2}}(s_0, s)) \sum_{a \in \mathcal{A}} q_{\pi_{\theta_1}}(a, s) \nabla \pi_{\theta_1}(a|s)
$$
$$
- p_{\pi_{\theta_2}}(s_0, s) \sum_{a \in \mathcal{A}} (q_{\pi_{\theta_2}}(a, s) \nabla \pi_{\theta_2}(a|s) - q_{\pi_{\theta_1}}(a, s) \nabla \pi_{\theta_1}(a|s)) \Big).
$$

We can further split the last sum into two parts:

$$
\sum_{a \in \mathcal{A}} (q_{\pi_{\theta_2}}(a, s) \nabla \pi_{\theta_2}(a|s) - q_{\pi_{\theta_1}}(a, s) \nabla \pi_{\theta_1}(a|s))
$$
$$
= \sum_{a \in \mathcal{A}} ((q_{\pi_{\theta_2}}(a, s) - q_{\pi_{\theta_1}}(a, s)) \nabla \pi_{\theta_2}(a|s) - q_{\pi_{\theta_1}}(a, s)(\nabla \pi_{\theta_1}(a|s) - \nabla \pi_{\theta_2}(a|s))).
$$

We can bound the norms of the terms as follows: for the first one, we have

$$
\Big| \sum_{s_0 \in \mathcal{S}} v(s_0) \sum_{s \in \mathcal{S}} (p_{\pi_{\theta_1}}(s_0, s) - p_{\pi_{\theta_2}}(s_0, s)) \sum_{a \in \mathcal{A}} q_{\pi_{\theta_1}}(a, s) \nabla \pi_{\theta_1}(a|s) \Big|
$$
$$
\leq \sum_{s_0 \in \mathcal{S}} v(s_0) \sum_{s \in \mathcal{S}} |p_{\pi_{\theta_1}}(s_0, s) - p_{\pi_{\theta_2}}(s_0, s)| \sum_{a \in \mathcal{A}} \|q_{\pi_{\theta_1}}\|_\infty \|\nabla \pi_{\theta_1}(a|s)\|_2.
$$

Using Proposition 11.4.3 and the inequalities $\|q_{\pi_{\theta_1}}\|_\infty \leq \frac{\|r\|_\infty}{1-\gamma}$ and $\|\nabla \pi_\theta\|_2 \leq 2C_\psi$, by (11.14) we see that the right-hand side is upper bounded by $C\|\theta_1 - \theta_2\|$ for some constant $C > 0$.

The other terms can be bounded in a similar fashion using Proposition 11.4.3; we leave the details to the reader. This shows that $\theta \mapsto \nabla J(\theta)$ is Lipschitz, which concludes the proof. □

Bibliography

[1] Robert J. Adler. *The Geometry of Random Fields*. SIAM, 2010.

[2] C. J. Albers, F. Critchley, and J. C. Gower. Quadratic minimisation problems in statistics. *Journal of Multivariate Analysis*, 102(3):698–713, 2011.

[3] Peter Bartlett, Yoav Freund, Wee Sun Lee, and Robert E. Schapire. Boosting the margin: A new explanation for the effectiveness of voting methods. *The annals of statistics*, 26(5):1651–1686, 1998.

[4] Mikhail Belkin, Daniel Hsu, Siyuan Ma, and Soumik Mandal. Reconciling modern machine-learning practice and the classical bias–variance trade-off. *Proceedings of the National Academy of Sciences*, 116(32):15849–15854, 2019.

[5] Mikhail Belkin, Daniel Hsu, Siyuan Ma, and Soumik Mandal. Reconciling modern machine-learning practice and the classical bias-variance trade-off. *Proceedings of the National Academy of Sciences*, 116(32):15849–15854, 2019.

[6] Vidmantas Bentkus. On Hoeffding's inequalities. *The Annals of Probability*, 32(2):1650–1673, 2004.

[7] James Bergstra, Rémi Bardenet, Yoshua Bengio, and Balázs Kégl. Algorithms for hyper-parameter optimization. In J. Shawe-Taylor, R. Zemel, P. Bartlett, F. Pereira, and K. Q. Weinberger, editors, *Advances in Neural Information Processing Systems*, volume 24. Curran Associates, Inc., 2011.

[8] Stephen Boyd and Lieven Vandenberghe. *Convex Optimization*. Cambridge University Press, 2004.

[9] Carmen Chicone. Introduction to ordinary differential equations. In *Ordinary Differential Equations with Applications*. Texts in Applied Mathematics, volume 34, pages 1–144. Springer New York, New York, NY, 2006.

[10] Lénaïc Chizat and Francis Bach. On the global convergence of gradient descent for over-parameterized models using optimal transport. In S. Bengio, H. Wallach, H. Larochelle, K. Grauman, N. Cesa-Bianchi, and R. Garnett, editors, *Advances in Neural Information Processing Systems*, volume 31. Curran Associates, Inc., 2018.

[11] Classifier comparison from scikit-learn. https://scikit-learn.org/stable/auto_examples/classification/plot_classifier_comparison.html. Accessed: 2023-10-30.

[12] G. Cybenko. Approximation by superpositions of a sigmoidal function. *Mathematics of Control, Signals and Systems*, 2(4):303–314, 1989.

[13] Amit Daniely, Roy Frostig, and Yoram Singer. Toward deeper understanding of neural networks: The power of initialization and a dual view on expressivity. In *Advances in Neural Information Processing Systems*, volume 29, 2016.

[14] Sanjoy Dasgupta. A cost function for similarity-based hierarchical clustering. In *Proceedings of the Forty-Eighth Annual ACM Symposium on Theory of Computing*, STOC '16, pages 118–127. Association for Computing Machinery, New York, NY, USA, 2016.

[15] Jacob Devlin, Ming-Wei Chang, Kenton Lee, and Kristina Toutanova. Bert: Pre-training of deep bidirectional transformers for language understanding. *arXiv preprint arXiv:1810.04805*, 2018.

[16] Harris Drucker. Improving regressors using boosting techniques. In *Proceedings of the Fourteenth International Conference on Machine Learning*, ICML '97, pages 107–115. Morgan Kaufmann Publishers Inc., San Francisco, CA, USA, 1997.

[17] Yoav Freund and Robert E. Schapire. A decision-theoretic generalization of on-line learning and an application to boosting. *Journal of Computer and System Sciences*, 55(1):119–139, 1997.

[18] Yoav Freund and Robert E. Schapire. A decision-theoretic generalization of on-line learning and an application to boosting. *Journal of Computer and System Sciences*, 55(1):119–139, 1997.

[19] Jerome Friedman, Trevor Hastie, and Robert Tibshirani. Additive logistic regression: a statistical view of boosting (With discussion and a rejoinder by the authors). *The Annals of Statistics*, 28(2):337–407, 2000.

[20] Jerome H. Friedman. Greedy function approximation: A gradient boosting machine. *The Annals of Statistics*, 29(5):1189–1232, 2001.

https://doi.org/10.1515/9783111288994-012

[21] Bolin Gao and Lacra Pavel. On the properties of the softmax function with application in game theory and reinforcement learning. *arXiv preprint arXiv:1704.00805*, 2017.

[22] Guillaume Garrigos and Robert M. Gower. Handbook of convergence theorems for (stochastic) gradient methods. *arXiv preprint arXiv:2301.11235*, 2023.

[23] Euhanna Ghadimi, Hamid Reza Feyzmahdavian, and Mikael Johansson. Global convergence of the heavy-ball method for convex optimization. In *2015 European Control Conference (ECC)*, pages 310–315. IEEE, 2015.

[24] Ian Goodfellow, Jean Pouget-Abadie, Mehdi Mirza, Bing Xu, David Warde-Farley, Sherjil Ozair, Aaron Courville, and Yoshua Bengio. Generative adversarial nets. In *Advances in Neural Information Processing Systems*, volume 27, 2014.

[25] Leonardo Ferreira Guilhoto. An overview of artificial neural networks for mathematicians, 2018.

[26] Soufiane Hayou, Arnaud Doucet, and Judith Rousseau. On the impact of the activation function on deep neural networks training. In *International Conference on Machine Learning*, pages 2672–2680. PMLR, 2019.

[27] Bobby He, Balaji Lakshminarayanan, and Yee Whye Teh. Bayesian deep ensembles via the neural tangent kernel. *Advances in Neural Information Processing Systems*, 33:1010–1022, 2020.

[28] Kurt Hornik, Maxwell Stinchcombe, and Halbert White. Multilayer feedforward networks are universal approximators. *Neural Networks*, 2(5):359–366, 1989.

[29] Arthur Jacot, Clément Hongler, and Franck Gabriel. Neural tangent kernel: Convergence and generalization in neural networks. In Samy Bengio, Hanna M. Wallach, Hugo Larochelle, Kristen Grauman, Nicolò Cesa-Bianchi, and Roman Garnett, editors, *NeurIPS*, pages 8580–8589, 2018.

[30] Motonobu Kanagawa, Philipp Hennig, Dino Sejdinovic, and Bharath K. Sriperumbudur. Gaussian processes and kernel methods: A review on connections and equivalences. *arXiv preprint arXiv:1807.02582*, 2018.

[31] Jason D. Lee, Ioannis Panageas, Georgios Piliouras, Max Simchowitz, Michael I. Jordan, and Benjamin Recht. First-order methods almost always avoid strict saddle points. *Mathematical Programming*, 176:311–337, 2019.

[32] Pierre Legendre and Louis Legendre. *Numerical Ecology*. Elsevier, 2012.

[33] Thomas M. Moerland, Joost Broekens, Aske Plaat, and Catholijn M. Jonker. Model-based reinforcement learning: A survey, 2022.

[34] G. Montavon, G. B. Orr, and K.-R. Müller. *Neural Networks: Tricks of the Trade*, 2nd edn. Springer Berlin, Heidelberg, 2012.

[35] Radford M. Neal. Priors for infinite networks. In *Bayesian Learning for Neural Networks*, pages 29–53. Springer, 1996.

[36] Ioannis Panageas, Georgios Piliouras, and Xiao Wang. First-order methods almost always avoid saddle points: The case of vanishing step-sizes. In *Advances in Neural Information Processing Systems*, volume 32, 2019.

[37] Samuel S. Schoenholz, Justin Gilmer, Surya Ganguli, and Jascha Sohl-Dickstein. Deep information propagation. In *International Conference on Learning Representations*, 2017.

[38] Bernhard Schölkopf, Ralf Herbrich, and Alex J. Smola. A generalized representer theorem. In *International Conference on Computational Learning Theory*, pages 416–426. Springer, 2001.

[39] Shai Shalev-Shwartz and Shai Ben-David. *Understanding Machine Learning – From Theory to Algorithms*. Cambridge University Press, 2014.

[40] LiWei Wang, XiaoCheng Deng, ZhaoXiang Jing, and JuFu Feng. Further results on the margin explanation of boosting: New algorithm and experiments. *Science China Information Sciences*, 55(7):1551–1562, 2012.

[41] Eric W. Weisstein. Modified bessel function of the second kind. From MathWorld – A Wolfram Web Resource. Last visited on 11/12/2023.

[42] Yale Face Database B. http://cvc.cs.yale.edu/cvc/projects/yalefacesB/yalefacesB.html. Accessed: 2023-10-30.

[43] Ge Yang and Samuel Schoenholz. Mean field residual networks: On the edge of chaos. In *Advances in Neural Information Processing Systems*, volume 30, 2017.

[44] Greg Yang. Scaling limits of wide neural networks with weight sharing: Gaussian process behavior, gradient independence, and neural tangent kernel derivation, 2019.

[45] Greg Yang and Edward J. Hu. Feature learning in infinite-width neural networks. *arXiv preprint arXiv:2011.14522*, 2020.

[46] T. Zeugmann. VC dimension. In C. Sammut and G. I. Webb, editors, *Encyclopedia of Machine Learning*. Springer, Boston, MA, 2011.

Index

https://doi.org/10.1515/9783111288994-013